Using Intelligent Automation for Shared Services Transformation

Dr. Vipin K Suri

December 2023

USING INTELLIGENT AUTOMATION FOR SHARED SERVICES TRANSFORMATION

Dr. Vipin K Suri

Managing Director, Shared Services International Inc.

ISBN: 978-1-962948-21-0 (Paperback)

978-1-962948-22-7 (Hardback)

ACKNOWLEDGEMENTS AND CONTRIBUTIONS

This book would not have become a reality without the encouragement and support of many individuals, including my daughter, Nina Arora, and my son-in-law, Pavan Arora.

I would also like to thank my daughter-in-law, Kirstin Suri, my son, Neil Suri and my wife, Purnima Suri, for their inspiration to author this book, which will serve as a practical guide for Shared Services organizations. The main purpose of authoring this book is to provide a resource for Shared Services organizations so that they can understand the benefits of digital technologies.

About the Author

Dr Vipin Suri is the Managing Director of Shared Services International Inc., a management consulting firm focused on Shared Services design and implementation as well as online and in-classroom training in the areas of Shared Services, Project Management and Six Sigma. He is also the Managing Director & Group CEO of SSI (Beijing), a management consulting firm in China. As a management consultant for over 21 years, Vipin has successfully assisted several companies in the Asia-Pacific Region, Australia, India, and North America with their efforts to implement Shared Services and to review the effectiveness of their Human Resources, Supply Chain Management, Finance, Information Technology, and other business support functions. Vipin's expertise areas include Finance, Accounting, Human Resources, Supply Chain Management, Real Estate & Facilities Management, Information Technology, Environment, Health & Safety, Strategic Sourcing, Procurement, Warehousing, Logistics, Mergers & Acquisitions, Electricity Distribution, Business Process Re-engineering, Outsourcing, Offshoring, Service Quality, Organizational Change Management, Training, Process Management, Governance Frameworks, Audit, Controllership, Customer Service and Sales.

Prior to becoming a management consultant in 2002, Vipin was Vice President of Shared Business Services for BHP Billiton Limited, one of the world's largest natural resources companies headquartered in Australia. He successfully achieved the desired outcomes ahead of schedule by leading efforts for designing & implementing Shared Services, implementing the SAP system within the company on a global basis, and re-engineering the hire to retire, order to cash, accounting to reporting, procure to pay and maintenance processes. Vipin established three multi-functional shared services centers for the company and outsourced business services for BHP Steel.

Before joining BHP in 2000, Vipin worked with Ontario Hydro, one of the largest electric utilities in North America. During his 26 years with the company, he worked in several senior positions, including General Manager - Mergers & Acquisitions, General Manager - Distribution Network Services and Asset Management, Vice President - Customer

Service, Vice President - Corporate Services and Vice President - Shared Services. He designed a model for and implemented Shared Services within the company with the establishment of service delivery centers. Vipin analyzed service delivery conditions for each service and outsourced those services that could not be delivered on a competitive basis. He redesigned the company's supply chain management processes and implemented profitable purchasing strategies by introducing the concepts of supply positioning, commodity management, strategic alliances, and life cycle cost management.

Vipin has published three books titled "Functional Automation and Digital Transformation," "The Silence of Solitude" and "Assessing Business Health of Shared Services." In addition, he has published several papers in Shared Services. Vipin has a doctorate (PhD) degree in Business Support Services (Shared Services). He holds a master's degree in business administration (MBA), a master's degree in mechanical engineering (M.Eng.) and a B.Sc. degree in Mechanical Engineering. He is a Professional Engineer (P.Eng.), a Chartered Professional Accountant (CPA) and a Certified Management Accountant (CMA). Vipin is also a member of the Association of Professional Engineers of Ontario, the Society of Chartered Professional Accountants of Ontario, the Society of Management Accountants of Ontario and the Worldwide Who's Who of Professionals.

Table of Contents

List of Figures

CHAPTER 1: INTRODUCTION

Intelligent Automation, or alternately Intelligent Process Automation, is a software term that refers to a combination of Artificial Intelligence and Robotic Process Automation. Companies use Intelligent Automation to cut costs by using Artificial Intelligence powered robotic software to replace workers who manage repetitive tasks. The term is hyper-automation, a concept identified by research group Gartner as being one of the top technology trends.

Intelligent Automation (IA), sometimes also called cognitive automation, is the use of automation technologies like Artificial Intelligence (AI), Business Process Management (BPM), and Robotic Process Automation (RPA) to streamline and scale decision-making across companies. Intelligent Automation simplifies processes, frees up resources and improves operational efficiencies, and has various applications in Shared Services organizations. Intelligent automation empowers rapid end-to-end business process automation and accelerates digital transformation. Digital technologies play a growing role in helping Shared Services organizations achieve their objectives. Companies have invested in Shared Services to reduce costs, develop capabilities, standardize processes, drive overall efficiency, and increase business value. There is a vast array of technological options that present a tremendous potential to completely revolutionize the way Shared Services operate.

Intelligence comprises reasoning, learning, problem-solving, perception and linguistic intelligence. The terms intellectual intelligence and emotional intelligence are well known, and now everyone is talking about artificial intelligence. When most people hear the term Artificial Intelligence, the first thing they usually think of is robots. In fact, artificial intelligence is based on the principle that human intelligence can be defined in a way that a machine can easily mimic and execute tasks, from the simplest to those that are even more complex. The goals of Artificial Intelligence include mimicking human cognitive activity. Researchers and developers in the field are making surprisingly rapid strides in mimicking activities such as learning, reasoning, and perception to the extent that these can be concretely defined. Some believe that innovators may soon be able to develop systems that exceed the capacity of humans to learn or reason out any subject. However, others remain skeptical because all cognitive activity is laced with value judgments subject to human experience.

As technology advances, previous benchmarks that define artificial intelligence become outdated. For example, machines that calculate basic functions or recognize text through optical character recognition are no longer considered to embody artificial intelligence since this function is now taken for granted as an inherent computer function. Artificial Intelligence is continuously evolving to benefit many different industries. Machines are wired using a cross-disciplinary approach based on mathematics, computer science, linguistics, psychology, and more. Algorithms often play a very important part in the structure of artificial intelligence, where simple algorithms are used in simple applications, while more complex ones help frame strong artificial intelligence.

Intelligent Automation is transforming Shared Services organizations by bringing innovative changes in how these organizations operate. The automation of business processes has reduced overall costs and increased the trust of internal customers. Automation is a necessary step towards transformation. Implementing and socializing automation across the entire Shared Service organization means that all teams benefit from the efficient workflow. Automating day-to-day workflows lies at the heart of successful transformation.

CHAPTER 2: AUTOMATION AND TRANSFORMATION

Service delivery models are always evolving. Companies are transforming to leverage global, multifunctional, and hybrid working models with tightening economic conditions. These companies prioritize areas across work, workforce, and workplace to succeed. Apart from expanding the traditional functional scope, they are accelerating digital capabilities like automation, analytics, and reporting to become more customer-centric. These Shared Services organizations are also promoting the social responsibility and diversity agenda and supporting a range of environmental, social, and governance processes. Shared Service organizations are taking the driver's seat in delivering transformational capabilities to the companies and are becoming more closely aligned to the company goals. Overall, what's clear is that Shared Services models are becoming more agile, digital, and cost-efficient as they seek to provide a better customer and employee experience.

The terms "automation" and "transformation" are often used interchangeably, but they have very different meanings. Automation involves applying technology to make processes run themselves, making those processes more efficient and increasing transparency and reporting capabilities. In short, automation is a means to an end, and the desired end is transformation. It can be argued that digital transformation requires not only making existing processes more efficient but also creating or unlocking new value for customers. It is true that customer expectations are constantly evolving with emerging technologies and that businesses are largely adopting digital transformation to continue to meet those expectations. However, this is a somewhat limited view of digital transformation, as it excludes any benefits of digital transformation that don't directly benefit the customer, such as transforming company culture by giving employees greater job satisfaction and allowing them to focus on more interesting projects than repetitive administrative tasks.

2.1 Automation

Automation is defined as a method that eliminates or reduces manual work. The concept of automation is well known; for many years, organizations have integrated automation as a change lever to improve operations. Advances in software and technology have made automation even more accessible to organizations. Automation is about changing the tasks that humans carry out so that they can focus on purposeful work, thus optimizing workplace productivity. Automation is one of the key capabilities supporting functional transformation. Organizations are transforming their functions to become more agile and efficient. This has been made possible as technology is playing an increasingly central role in driving these changes through the process of digital transformation. Automation is not about replacing humans with robots. The best results are achieved when humans work alongside Robots and Artificial Intelligence to enhance complementary capabilities. Robots or 'digital workers' work best when they perform transactional, data-intense, and highly repetitive tasks, allowing people to focus on the more innovative, creative, and strategic tasks.

From a business standpoint, eliminating the need for manual intervention in complex or repetitive business processes can have a positive impact on operations. By automating certain decisions, organizations can accelerate process workflows and improve the accuracy and consistency of the results. Business process automation is the use of technology to execute recurring tasks or processes in a business where manual effort can be replaced. It is done to achieve cost minimization, greater efficiency, and streamlined processes. Digital process automation is fundamental to delivering better customer experiences that fully meet user expectations. However, it is also part of a broader shift towards using business objectives and requirements to design enterprise services versus relying on default system configurations that constrain performance.

Automation has streamlined and increased the ability of Shared Services organizations to scale and perform more work both faster and better without relying on additional employee headcount. In fact, many organizations are moving from high-touch, manual processes to digital operations. Automation incorporates a host of tools, from robotic process automation to artificial intelligence. These tools improve employee productivity, enhance decision-making, and help optimize systems. Automation has expanded beyond simply achieving gains in cost and efficiency to encompass a customer-centric focus on improving user experience.

There are few fields of organizational activity that are untouched by automation. Automation has been helping businesses in every industry around the world seek efficiencies that come from replacing manual tasks with machine-operated ones. Many Shared Services organizations have aimed to adopt automation technologies for business growth and to encourage greater creativity, productivity, and innovation from employees. Automation involves machines, including computers, that can complete repetitive tasks at a fraction of the time it takes humans and to do so with greater accuracy and without complaint.

Automation is impacting employees in a variety of complex ways, so it is critical for employees to have a greater understanding of and training on different software systems to ensure they keep up with the automation and benefit from it rather than viewing it as a potential threat to their jobs. There is no way to stop technology. As computers take over the processes, people who simply process transactions could be offered new roles where they identify the value in information and help make decisions based on that information. Automation is critical to managing, changing, and adapting functions and how organizations operate through processes. By simplifying change through automation, the Shared Services leaders gain the time and energy to focus on innovation. The automated enterprise's goal is to get work done faster.

Intelligent automation (IA) is comprised of three cognitive technologies. The integration of these components creates a solution that powers business and technology transformation (What is intelligent automation?, 2020).

1. The most critical component of intelligent automation is artificial intelligence or AI. By using machine learning and complex algorithms to analyze structured and unstructured data, businesses can develop a knowledge base and formulate predictions based on that data. This is the decision engine of IA.

2. The second component of intelligent automation is business process management (BPM), also known as business workflow automation. Business process management automates workflows to provide greater agility and consistency to business processes. Business

process management is used across most industries to streamline processes and improve interactions and engagement.

3. The third component of IA is robotic process automation (RPA). Robotic process automation uses software robots, or bots, to complete back-office tasks, such as extracting data or filling out forms. These bots complement artificial intelligence well, as RPA can leverage AI insights to handle more complex tasks.

Integrating these three components creates a transformative solution that streamlines processes and simplifies workflows to improve the customer experience. Intelligent automation platforms provide many benefits across industries because of the use of large amounts of data, precision of calculations, analysis and resulting business implementation. Key benefits include:

- **Reduce costs by augmenting the workforce and improving productivity:** Automating systems and processes and using data and analysis to ensure accuracy can accelerate production. IA also provides the ability to scale quickly without increasing risk, compromising quality, or putting a strain on the existing workforce. Business leaders are reaping the benefits of this capability through higher production yields and improved ROI.
- **Improve accuracy through consistent processes and approaches, which enhances quality:** The strength behind intelligent automation is the use of artificial intelligence to drive decision-making and bring a consistent approach to repetitive tasks.
- **Improve customer experience:** Providing a higher quality, more reliable product to market faster or getting answers to queries faster (or immediately) provides a richer, more positive experience for the customer and, thereby, a competitive advantage for the company.
- **Address compliance and regulations with confidence:** Many industries are required to adhere to specific regulatory policies. Intelligent automation can leverage its task automation features to prove a more consistent approach to compliance.

Intelligent automation streamlines processes that were otherwise comprised of manual tasks or based on legacy systems, which can be resource-intensive, costly, and prone to human error—the applications of IA span across industries, providing efficiencies in different areas of the business.

2.2 Transformation

Business transformation is an umbrella term for making fundamental changes in how a business or an organization runs. This includes personnel, processes, and technology. These transformations help organizations compete more effectively, become more efficient, or make a wholesale strategic pivot. Corporate and business support functions such as Finance, Human Resources, IT, Procurement, Supply Chain Management, Legal, Real Estate, Facilities Management, etc., represent a significant investment for companies and are essential to almost every company. These functions provide business services required by various units within a company to support their operations and meet their strategic business objectives. These services are invisible to the external customers of the company but are critical for running the day- to-day business, e.g., payment of employee wages, payment of supplier invoices, maintenance & support of application systems, recruitment, training & development, tax advice, etc. As these services are primarily support in nature, the fundamental principle in offering these services is to improve efficiency and effectiveness and thereby reduce costs and increase value offering, e.g., delivering services at higher quality and lower cost than the business units can achieve individually. But these functions also represent an enormous transformation opportunity in an environment of increased globalization, technology disruption, and cost pressures.

Digital transformation is the practice of using technology to radically change your business to pursue new revenue streams and business models and to avoid disruption. It involves more than simply moving everything to the cloud and adopting a variety of digital processes. It radically rethinks how an organization uses technology, people, and processes to fundamentally change business performance. Digital transformation is not an isolated IT project. Instead, it is the combination of numerous projects that transform every aspect of a company.

Some of the ways this can play out are as follows:

- Upgrading underlying technology (renovating or replacing legacy systems),
- Shifting the way the customers are served (increasingly via mobile or across multiple different channels),
- Automating processes for greater efficiency and
- Integrating with third parties to deliver better customer experiences.

True digital transformation is about leveraging all the digital tools, including automation and data analytics capabilities, to transform how a company operates and makes decisions fundamentally. This includes fostering a culture of innovation and continuously iterating business models to deliver the best products and services. Successful digital transformation is proactive instead of responsive. To stand a chance of doing the same thing in companies, we need to have fingers on the pulse of the company and on the customer relationships, for which the insight offered by data analytics is indispensable. Figure 2.1 shows a conceptual framework for digital transformation.

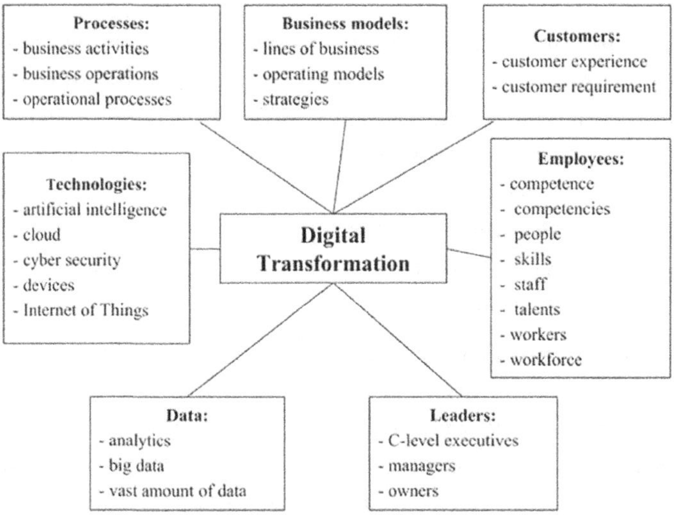

Figure 2.1: Conceptual Framework for Digital Transformation (Semantic Scholar)

Functional organizations in major companies are facing dramatic change. Businesses are becoming more focused on capabilities and demanding that leaders of functions play a more strategic role. With less stable environments and more intense competition in many markets, there is a premium on agility and flexibility and an increasing emphasis on performance results. These pressures are forcing many functions to boost operational excellence even further, usually while reducing costs.

The core elements of digital transformation are shown in Figure 2.2.

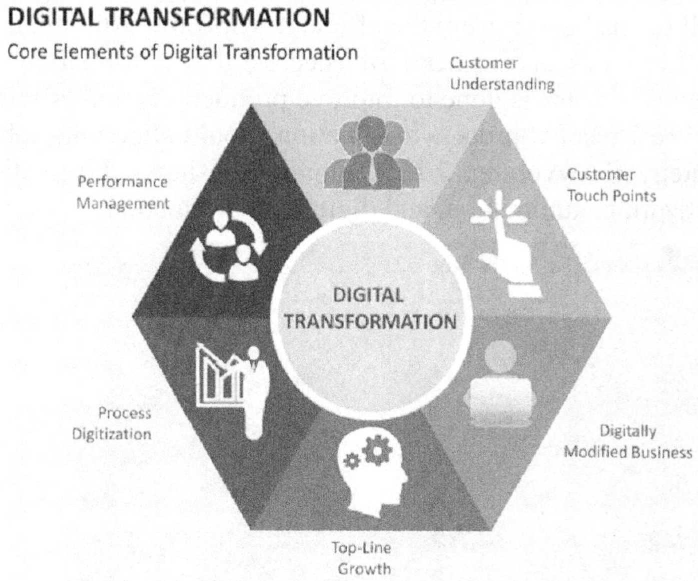

Figure 2.2: Core Elements of Digital Transformation (Bithgroup Technologies)

Transformation takes place at dizzying speeds, requiring a level of integration and alignment that many companies are not prepared to manage. Because many initiatives that once took five or six years to complete are now being implemented in a year or less, many organizations need to manage mountains of information, compressed times, and countless decisions affecting all aspects of strategy and operations. The risk of failure is great, but companies that can pull the disparate elements into alignment can achieve bold objectives, now and into the future. Transformation is not about one massive storm sweeping every long-standing tradition away in one blow. It is mostly a long journey of slight changes that happen every single day in bits and pieces. That is why consistency and willingness to repeat and iterate are so important in transformation.

2.3 Automation: Key to Transformation

Automation is a necessary step towards transformation. Implementing and socializing automation across the entire Shared Service organization means that all teams benefit from the efficient workflow. Automating day-to-day workflows lies at the heart of successful transformation. While optimizing the way work is done to improve productivity and efficiency is desirable, for best results, process automation should align with a broader business strategy (Frevvoblog, 2022). Figure 2.3 shows the relationship between digitization, automation, and digital transformation.

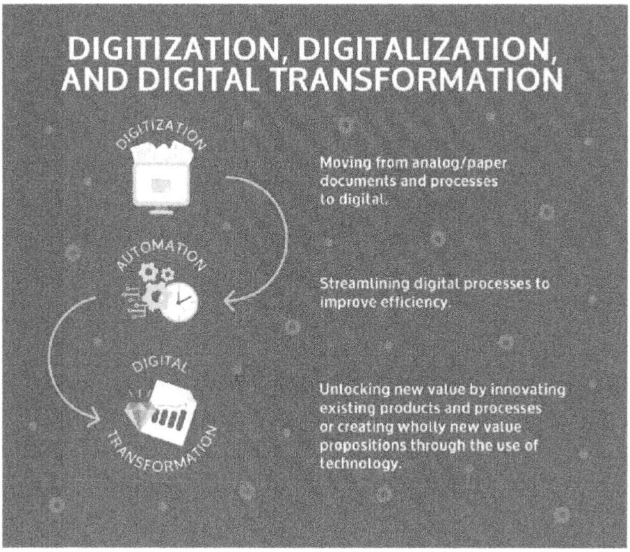

Figure 2.3: Digitization, Automation and Digital Transformation

Process automation is a necessary step towards digital transformation. Implementing and socializing automation across the entire company means that entire teams benefit from newly efficient workflow, not just IT. Automating day-to-day workflows lies at the heart of successful digital transformation and is useful to digitize forms for employees and customers, automate routine processes, and measure their effectiveness. With this data, we can reduce pain points to drive genuine change and improve business performance in myriad ways. Without workflow automation, we will still be dealing with manual processes, in which case the threat of being disrupted and becoming irrelevant is real. While optimizing the way work is done to improve productivity and efficiency is desirable, for best results, process automation should align with a broader business strategy. Here are a few ways automation can help set the path to transformation (Frevvoblog, 2022):

- **Deliver World Class Customer Service**

  Over-delivering customer service is the single most important thing we can do to increase revenue. Customers are accustomed to being able to interact with Shared Services organizations online at any time. They will be frustrated and feel inconvenienced if they need some information that is only available during business hours. Automation ensures that customers have a smooth experience by enabling them to access information when they need it. When employees are wasting time performing manual tasks like chasing down internal approvals, they have less time to focus on providing excellent customer service, whereas automated approvals free up their time so it can be invested into what's important: to customers.

- **Increase Efficiency and Accuracy so Employees Can Focus on Work That Matters**

  Automation can accelerate operations significantly while also reducing errors by taking over all the repetitive manual tasks previously performed by people. With increased efficiency comes increased capacity, making it easier to scale operations. Undoubtedly, the following applications might be transformed:

- Automate forms and workflows so there is no need to print, sign and scan.
- Integrate systems and databases to reduce tedious manual data entry.
- Intelligently route documents to the right person at the right time to ensure that business constraints are satisfied.
- Automatically support all mobile devices and speed up the approval process by providing 24/7 access from anywhere.
- **More Rewarding Work, Better Workplace Culture**

  Automation frees people to perform more challenging or creative tasks, which is ultimately more rewarding and makes for more satisfied and engaged employees. When the staff is engaged, they care more about the work they do, which translates into success.

- **Facilitate Teamwork by Breaking Down Silos**

  Integrated automated workflows can bridge gaps and break down silos to promote collaboration between different teams, facilitating cross-functional cooperation. Exposing the employees to different functions allows them to get greater insight into the business, creating a climate favorable to innovation. With process automation, we can automate workflows between business support functions.

- **Improve Transparency and Strategic Oversight**

  When we automate business processes, we gain access to an abundance of data that we didn't necessarily track before. This offers increased visibility into which processes are successful and which ones could be further optimized. The analysis of this data allows for better strategic positioning and risk assessment, allowing the development of value propositions that live up to the promise of transformation.

- **Boost Security and Compliance**

  Automation can contribute to security and compliance in several ways. Improving security and compliance were primary drivers for implementing automation tools. Automation makes identifying vulnerabilities easier and rolling out fixes, empowering organizations to be proactive and address vulnerabilities before breaches occur.

- **Cut Costs**

  By automating processes like data entry, document management, finance processes, and approval workflows, we can save not only time but also money by reducing waste._While automation does not equate to transformation, it's a pivotal part of the process.

- **Meet Customers at their Preferred Channel**

  Customer-centric companies are focusing on the mobile customer experience and have ramped up investment in this area. It is crucial that customers can reach the business via mobile regardless of the mobile browser they use. It is increasingly easy to make your website mobile responsive, and the workflow automation systems allow you to create forms that are already optimized for mobile display.

- **Make the Business More Agile**

  Automation empowers an organization to react rapidly to changing business conditions. Consider two key capabilities:

- **Responsiveness**

Enable employees to do the right thing at the right time. This means reducing business complexity and promoting collaboration (access to people and information from anywhere, as needed).

- **New Opportunities**

  Faster rollout of new business initiatives as business conditions change, e.g., introducing a new product offering or modifying internal processes to quickly satisfy changed customer demands – even something as simple as enabling customers to confirm their identity by taking a selfie while holding an identity document.

  Both capabilities offer a significant competitive advantage, which is, after all, the aim of transformation.

2.4 Business Workflows to Automate for Transformation Success

Automation can play a key role in the transformation journey by streamlining and optimizing operations, promoting collaboration, granting greater oversight, and generally helping to create a climate conducive to innovation. The following processes can be considered for automation to set the Shared Services organization up for transformation:

- **Onboarding**

  Automating the employee onboarding processes can profoundly impact the ease and speed with which an organization can bring them into the fold, which plays an important role in making a good first impression.

- **Finance Processes**

  Automating Finance processes such as accounts payable and accounts receivable can improve the cash flow, but more importantly, it frees up the Finance team's time to allow them to do the analytical work. Finance professionals can play a vital role in business transformation, so it is really in the company's best interest to improve the Finance processes.

- **HR Processes**

  Streamlining HR processes makes HR's job easier and saves time, which can be used to attract and retain world-class talent. Not only that, but automation can also ensure compliance with regulations and improve data privacy and security, which is essential when dealing with sensitive information. Moreover, with interdepartmental workflow integration, orchestrating tasks like training and certification becomes much easier.

- **Approval Routing**

  Processes that involve multiple approval stages can be simplified by using software to ensure that nobody needs to run around chasing down supervisors for signatures. Automated reminders can reduce bottlenecks and notifications help to keep various parties in the loop about the status of the request, while responsive forms make it easy for supervisors to respond to requests on the go.

2.5 Transformation Planning

Transformation planning is developing a strategic plan for modifying an organization's business processes through modifying policies, procedures, and processes to move the organization from the current state to the future state. Business transformation is the opportunity to define a bold ambition that goes beyond incremental change, the opportunity to rethink business and operating models to deliver breakthrough value. It involves strategic decisions that affect where the organization will grow, how it will operate, and what performance improvements will be expected. The Finance, HR and IT functions have evolved from playing a supporting functional role to becoming a key player in determining and driving corporate strategy.

The digital future of business support functions is exciting. As traditional, time-intensive tasks like forecasting, reporting, and distributing information come to be handled through smart digital tools, people can focus on asking the right questions, running and interpreting analyses, drawing conclusions, managing tradeoffs, and generating insights that help business units create new value. Capitalizing on this potential requires that companies start putting the right organizational structure in place. An effective business transformation plan can help drive lasting positive change in people and their work through better leadership, strategy, and culture.

Change Management is the process of obtaining the intelligence to perform transformation planning by assessing an organization's people and cultures to determine how changes in business strategies, organizational design, organizational structures, processes, and technology systems will impact the company. Business transformation refers to medium and large-scale projects that impact strategies, management systems or business processes. Large-scale transformations may result from the acquisition of a new company, market disruption, business model innovation, restructuring for revenue growth or performance improvement and operational change. Once the need for change is recognized, the transformation journey can begin. The steps outlined below can serve as a guide for Shared Services organizations undertaking business transformation:

1. **Strategy Development:** A simple approach to strategy for business transformation includes components such as Mission Statement, Vision Statement, Values, and guiding Principles. The strategy establishes the initial parameters governing the business transformation.
2. **Current State Assessment:** To effectively pursue the transformation of a business, it is critical to begin with an understanding of the current state of its operations. Current State Assessment includes developing an understanding of the current strategy, measuring baselines for key performance metrics, mapping various business processes, outlining the staff's existing skill sets, and documenting the capabilities of technology and information systems. Successful support of individuals and organizations requires a transition from the current to the future state. Conducting a current state assessment provides strategic insights into the complexity of the impact of the change on the organization. Typically, the content or focus of the assessment comes from the insights gained by conducting a gap analysis between the current

state of the organization and the future state defined through the strategy and vision for the transformation program.

3. **Organizational Change Management:** Given that a business transformation is by nature a transformational change for the organization, it is critical to consider and plan for its impact on all aspects of the business and its employees. Organizational Change Management focuses on making employees aware of and engaged in the change effort.
4. By defining and completing a change process, an organization can better define and document the activities that must be managed. Moving through these stages will help ensure effective, long-term, and sustainable results. These stages unfold as an organization moves through the transition phase in which the required transformational changes are introduced, tested, understood, and accepted in a manner that enables individuals to let go of their existing behaviors and attitudes and develop any new skills needed to sustain desired business outcomes. Executed properly, OCM can remove obstructions that prevent change from occurring, accelerating the transformation and positively impacting post-transformation sustainment.
5. **Implementation and Change Resistance Plan Development:** In concert with the Organizational Change Management Plan, develop an Implementation plan to make the aspirational vision a reality. The implementation plan should include steps to address resistance, which is a critical element of organizational change activities. Resistance may be a unifying organizational force that resolves the tension between conflicts that are occurring as the result of organizational change. Resistance occurs as the unit or individual perceives how the change will affect its likelihood of voicing ideas about organizational change. Resistance is often seen as a negative force during transformation projects. However, properly understood, it is a positive and integrative force to be leveraged. It catalyzes the converging and diverging currents between change leaders and respondents and creates agreement within an organizational system.
6. **Implementation Plan Execution:** Rushing implementation can place extreme pressure on existing operations and staff, resulting in resistance to change. However, the implementation of changes is the last step of a business transformation initiative and the risks associated with implementation need to be managed.

7. **Post-Implementation Support:** Post-implementation support is necessary as success is ensured by embedding and continually evolving the operating model.

A key component of the transformation plan should address the strategic communications required to support the implementation of the transformation. Open and frequent communication is essential to effective change management. When impacted individuals receive the information (directly and indirectly) they need about the benefits and impact of the change, they will more readily accept and support it. Organizations need to consider the following areas to deliver business transformation projects successfully:

- Clearly understand the problem.
- Communicate the change.
- Develop a business case.
- Balance clear vision with detailed design.
- Start early and move fast.
- Understand the baseline costs.
- Organize the retained functions.
- Prepare for digital workforce management.
- Develop a digital change management program.

It is important to maintain a set of working guidelines for the organization to provide stability and resilience. Using an operating system that integrates people, processes and technology will support and guide employees, keeping constant focus during the implementation of the new strategy and its associated dynamic action plan. Integrating people, processes and technology is essential for the efficient running of the business. This is achieved by ensuring that people are engaged and committed, processes are efficient and effective, and applications are used effectively, and service disruptions are prevented.

2.6 Conclusion

Business transformation is a change management strategy that can be defined as any shift, realignment, or fundamental change in business operations. The aim is to make changes to processes, people, and systems to better align the company with its business strategy and vision.

Commitment to completing the change process is vital to a successful outcome. There are many reasons a business might consider a transformation, for instance, new technologies, shifts within the market, low profit and turnover, or a merger and acquisition. It may involve changes to the entire organization, such as when integrating the two companies involved in a merger or acquisition. Or it may involve a change to a specific function, such as IT, HR, or Finance. For example, a drive for greater efficiency may prompt the implementation of a new Finance system.

Internet of Things, AI, and robotics are seen by global organizations as both the most disruptive technologies and the most important for cutting costs. Organizations cannot fall back on technology gaps as an excuse to delay implementing other elements of a transformation, but at the same time, some organizations have found that they simply cannot move forward on a transformation until technology has caught up. The use of technology is integral to everything we do at work and outside the workplace and support functions need to have a clear road map for how they see their IT architecture evolving over time. Embracing the opportunity to transform corporate and business support functions fundamentally requires strong leadership across the organization, good communication of direction, and not being too hasty in trying to unlock immediate benefits. New technologies allow support functions to work in new ways and to streamline their overall operating model. Specific enabling technologies include the following:

- **Big Data.** New tools allow companies to collect, validate, store, and process large amounts of information - at far lower cost and higher quality levels than in the past—and make it more accessible to leaders. For example, data lakes can work with the existing unstructured data that many organizations already have (often in tremendous volume) but do not yet capitalize on. As a result, support functions can proactively identify business needs and even develop solutions—through increased transparency and smarter business insights.
- **Global Connectivity.** Cloud computing, software-as-a-service, and mobile/wearable devices make data-driven insights more accessible throughout the organization in real-time. This increased connectivity allows support function employees to collaborate in new ways and make smarter and faster decisions.

- Artificial Intelligence and Machine Learning. Advanced technologies allow companies to automate repetitive transactional tasks and rule-based activities through robotic process automation. And with machine learning, technology can increasingly handle processes that require some judgment as well. Algorithms can work 24/7 with greater accuracy and at lower cost. Even self-service functions become smarter through AI.

Dynamic and challenging markets, a changing regulatory landscape, and intense pressure on cost containment are just some of the issues that have contributed to the elevation of support functions as a strategic corporate force. There are three different categories of transformation, namely operational, operating model and strategic. The operational category is doing what is currently being done but faster, cheaper, and better. The operational category involves doing what is currently being done but in a fundamentally different way. The final category is strategic, which is the most promising and involves changing the very essence of a company. Strategic transformation changes a company's competitive set. Leaders should be thinking about how to blend operational models and strategic transformation to connect and coordinate through carefully constructed capabilities to rise to the challenge of disruptive change and their future.

Automation and digital transformation have become two of IT's dynamic duos. Where you see or hear one, the other seems sure to follow. They are certainly linked by their priority and popularity in IT and business circles. Automation and digital transformation have been atop the strategic roadmaps of IT leaders and other leaders. Digital transformation has become an industry unto itself. What is the actual relationship between automation and digital transformation? Answers to questions like this require more context and nuance. Let's ground the relationship in clear definitions. Here's how we can describe digital transformation while also recognizing that any company's unique needs and characteristics mean that the digital transformation must necessarily be flexible.

Digital transformation is the integration of digital technology into all areas of a business, fundamentally changing how it operates and delivers value to customers. It is also a cultural change that requires companies to continually challenge the status quo, experiment, and get comfortable with failure. It is the scope and scale of that definitional change and the value it aims to produce that brings to automation. Automation is one of the major catalysts that make it possible to execute digital transformation goals.

CHAPTER 3: SHARED SERVICES

Shared services are when a company combines its business support functions to better serve the company and its business units, running as a business within the company utilizing a well-defined infrastructure to enable higher value service delivery. Typically, this includes services such as human resources, finance and accounting, information technology, supply chain, and other front and back-office services.

By combining these services into a centrally managed organization, businesses can achieve efficiency and cost savings that would otherwise be difficult to achieve through decentralized operations. In addition to helping the company, shared services also provide employees with valuable career opportunities and training (Scottmadden, 2022).

3.1 Shared Services Concept

Shared Services represent a strategy for increasing the efficiency and effectiveness of business support functions in companies:

- **"Shared"** means **service providers** and **customers** sharing responsibility for results attained. **Shared** also means sharing resources required to deliver services to various business units in a company.
- **"Service"** means **work performed and outputs produced** that deliver results to **internal customers,** e.g., payment of suppliers is a service, whereas verification of invoices is an activity and not a service.

Shared services or Shared Services Centres refer to a dedicated unit (including people, processes, and technologies) structured as a centralized point of service and focused on defined business support functions. These functions are supported by information technology and IT services for multiple business units within the company. Companies globally have been deploying the shared services delivery model. Shared services continue to be attractive because they hold the potential for:

- Faster decision-making, more informed policy-making, more effective workforce management and improved resource alignment with the company goals.
- Improved servicing ration/response times, reduced cycle times and improved automated reporting.
- Reduced duplicative software/hardware/operations/labor resources.
- Increased interactivity with constituents, including improved communication and responsiveness.
- Enhanced quality, timeliness, accuracy, and consistency.

However, these goals are not always realized because even though the functions seem duplicative and logical candidates for shared services, they are not always severable and/or have complicating factors.

3.2 Shared Services Model

Shared Services are best described as the consolidation of business support operations that are used by multiple areas of the same company. A Shared Services model is considered cost-efficient because it can centralize back-office operations that are typically used by many different divisions of the same company and eliminate operational redundancy. Shared services need to have the mindset of a fully optimized business and view the rest of the company as internal customers. As an organization built around delivering services, these people are accountable for balancing the cost of delivering value while maintaining strong service levels. Alongside this, they are tasked with identifying ways to use the operating model.

This operating model is constructed around three crucial components: people, processes, and technology. Even though the centralization of services is an important part of shared services, the focus is also on a much broader goal. Shared services should help to gain efficiencies far beyond the consolidation of operations - promoting a culture of continuous improvement. How does a shared service business model work? Are you trying to build a request-centric operations model that places an emphasis on simplified customer experience?

There has never been a better time to start exploring the consolidation of business support services. A smooth operational structure often simplifies the best customer experiences. Taking a centralized approach to operations can often be more cost-efficient than giving the user multiple points of access and lead to fewer headaches for those managing daily requests. When multiple divisions in the same company run separately, it is not always easy for employees to navigate around the various divisions to get their solutions fulfilled. A general lack of organization from the service providers can result in a convoluted internal experience. Fortunately, a shared services model will bring order to the daily operation.

To understand the Shared Services Model, the roles and responsibilities of the Corporate Center, Strategic Business Units and Shared Services Organization should be clarified as follows:

- The Corporate Center exists to guide and enhance the efforts of the Strategic Business Units to maximize returns to shareholders. The Center:
- Ensures compliance with all legal and regulatory requirements for which the company is liable,
- Establishes policies on behalf of the Company that all Strategic Business Units must comply with,
- Strategic Business Units (SBUs) exist to generate accounting profit and positive cash flow by designing, manufacturing, and marketing products and services into global markets. Each Strategic Business Unit:
- Owns assets and seeks to generate an acceptable return on those assets,
- Makes all decisions about its own business, within boundaries established by legislative and regulatory authorities, and the Corporate Center,

- Purchases business support services from Shared Services according to specifications and for a price jointly established between the Strategic Business Unit and Shared Services.
- The Shared Services organization exists to provide services required by the Strategic Business Units and the Corporate Center to fulfill their respective missions and:
- Provides only those services that its customers (Strategic Business Units and the Corporate Center) at a level that those customers agree to pay for,
- Competes for the business it gets from its customers,
- May provide services that assist SBUs in meeting their obligations to legal and regulatory authorities and to the Corporate Center.

Figure 3.1 shows elements of the Shared Services Delivery Model.

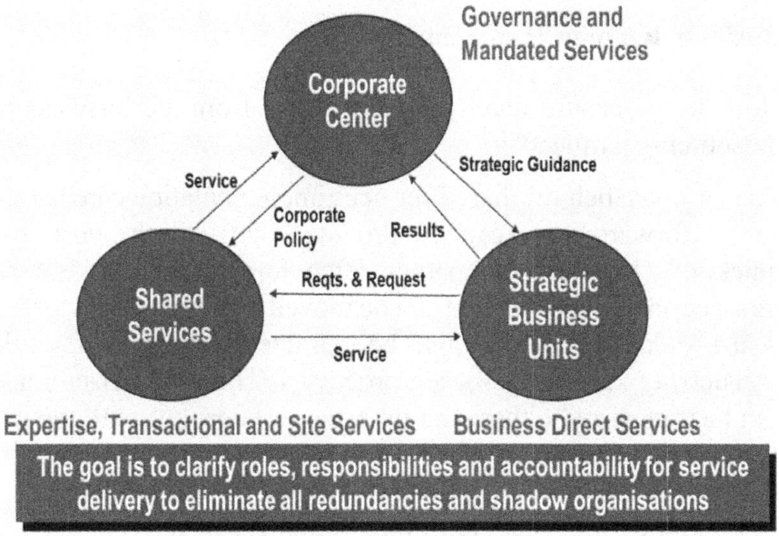

Figure 3.1: Shared Services Delivery Model

Shared Services Leaders have a complex role. The challenges and opportunities to get everything right are as follows (Howard, 2022):

- **Process Execution**

  How do we minimize errors, increase speed, drive down transaction processing costs, and adapt quickly to support new in-kind transactions and revenue streams?

- **Process Improvement**

  How do we make wise decisions on technology to solve the most impactful process problems and support changes in working capital needs as the supply chain (vendors and customers) resets itself? Changing a process can lead to great efficiency gains, but the consequences for other areas of the business that are not directly involved in the change can be profound. Unless these consequences are fully analyzed and possible dangers understood and mitigated, the efficiency gains of process improvement could pale in comparison to the costs of disruption to operations across the organization. Transforming a process in isolation is a sure-fire way to create difficulties elsewhere in the organization. Just as "no man is an island," nor is any business process.

- **Business Partner Management**

  How do we ensure quality in processing from the business process outsourcing partners?

  The rationale behind shared services implementation can be relatively straightforward: "we can do Process X at half the cost, twice as quickly." However, companies often forget that it is not just the transactional processes that can be moved into shared services; a great deal of added value can also be extracted from shared services by transferring knowledge-based processes. These higher-end activities can be transferred either straight away or later (but with the potential to do so included in the plan). After all, setting up shared services is a significant cash investment and a change management effort, and it is worth getting the most out of the investment by moving over work.

- **Process Governance**

  How do we mitigate risk and comply with ever-changing, country-specific regulations for workforces, data privacy, statutory reporting, tax rules, document retention policies and more?

- **Employee Engagement**

  How do you keep a diverse and remote workforce located in multiple centers around the world engaged and productive?

- **Customer Service**

  How do we provide prompt customer service to customers in multiple geographies, time zones and languages?

- **Customer Relationship Management**

  How do we ensure that the diverse customers are getting what they need?

- **General Management**

  How do we achieve year-over-year cost reduction goals while doing more and being positioned for future growth?

3.3 Shared Services Advantages

When it comes to business units, there are many advantages to implementing a shared services delivery model:

- **Cost Reduction:** Perhaps the most obvious benefit is that shared services can help to save time and money. By combining various functions into one service provider organization, business units can often achieve economies of scale and reduce duplication of efforts.
- **Improved Service:** Centralizing certain functions can make it easier to check and manage performance. When services are siloed or distributed within business units, it can be difficult to track progress and find areas for improvement. Standardized processes and the use of service management tools help improve service.
- **Better Controls and Insights:** Shared services enable data centralization, which supports improved reporting and analytics. It

also helps to improve compliance and enable better decision-making by the company.

- **Improved Communication:** Shared services can improve communication and collaboration between departments and employees by breaking down silos and encouraging information sharing.
- **Greater Flexibility:** Shared services can give companies the flexibility to scale up or down as needed without having to make significant changes to their organizational structure. During mergers and acquisitions, shared services can simplify the process of integrating companies to combine operations and achieve synergies.
- **Improved Productivity:** When everyone works from the same platform or uses the same service provider, it can make sharing information and working together on projects much easier.

3.4 Shared Services Goals

Technology has often been the driver for shared services within a company because it can be expensive to buy, keep and train employees to use. By combining their resources, the business units reap full advantages of technologies that might be too costly if bought on their own while realizing efficiencies in staffing and improved customer service through collaboration. The shared services strategy allows business units to focus more on their core competencies and strategic priorities rather than business support services that are critical but serve rather than drive their missions.

Shared Services is not a single idea or solution; it is a set of ideas that provide an:

- Innovative way of looking at work performed by staff in business support functions.
- Understanding of the needs for and costs of all work performed.
- Approach for redesigning services cost-effectively.
- Experientially based process for transitioning staff from being ***provider-driven*** to becoming ***customer-focused.***

As such, it addresses two important and related dimensions.

- The "what's," or the services that need to be performed and how they are organized to leverage delivery optimally.
- The "how's," or the process for implementing the agreed-upon model for effecting the desired results and true customer service behavior.

The **"what"** goals are to use the delivery of all services **to *reduce cost and maximize productivity, customer satisfaction, quality, and resource configuration concurrently by:***

- Supplying only agreed-upon services at the "right" levels to satisfy internal customer requirements.
- Achieving performance standards acceptable to customers at competitive rates through best practices

The **"how"** goals are to help shared service staff reach and sustain success by becoming ***service-oriented, business-based contributors.***

Service-oriented means that they know the

- Services they deliver.
- Customers (users, payers, and stakeholders) served at the service level.
- Expectations, quality attributes and relative importance of each service.

Business-based means that they know the

- Total service delivery costs (Staff and Non-staff).
- Staff time and resources are spent on a service basis.
- Productivity and performance.
- External comparative rates of alternative service providers.

Contributors mean that employees have the requisite:

- Service orientation towards internal customers.
- Technical competencies to perform their work.
- Value commitment to the vision and goals set forth.

The **"whats"** are easier than the **"hows,"** especially for staff who:

- Have progressed through traditional career ladders.

- Are not familiar with alternative approaches.

3.5 Conclusion

The types of activities performed by Shared Services organizations span from transaction-based, that is, routine, high volume activities, to transformation-based, that is, activities that require extensive expertise and are strategic to the company. The business support functions that may be shared are very diverse, including both front-office work, such as customer support, and back-office work, such as finance, law, human resources, and information systems (Ulrich, 1995)

Shared Services effectiveness is dependent on the methodology used and the service model deployed. Whether the shared services initiative takes place in finance, human resources, facilities, legal services, or some other part of the support area, the goal is to better leverage resources to create value, reduce costs, and increase the quality of support work. The liberated resources are not simply added to profits. In many organizations, the savings from shared services are being reinvested in increasing the value-added content of the company's products and services, resulting in competitive gains and revenue/market share growth. Shared services are not a panacea or fad - these are a practical solution to the longstanding need to control nonessential costs in a competitive environment.

CHAPTER 4: INTELLIGENT AUTOMATION

The pace of change is accelerating across Shared Services organizations. There has been a shift from service provision to value-add. There is a growing expectation from the internal customers that Shared Services teams should deliver added value to the company. Increasingly, Shared Services organizations are expected to drive greater efficiency in the company while also enhancing client experiences and relationships. Shared Services leaders see the benefits of Intelligent Automation (IA) to improve quality and productivity while empowering employees to focus on higher-value work. A digital workforce powered by intelligent automation is the answer to managing costs simultaneously, maintaining controls and compliance, and meeting high delivery expectations. As a result, leading companies have been able to transform their Shared Services organizations through Intelligent Automation.

Intelligent Automation (IA) is a combination of Robotic Process Automation (RPA) and Artificial Intelligence (AI) technologies, which together empower rapid end-to-end business process automation and accelerate digital transformation. Intelligent Automation is a term that describes a holistic solution for digital transformation, mainly based on Business Process Management to orchestrate users, tasks and systems depending on the business needs at each moment. It also contemplates the use of analytics and Artificial Intelligence to make automated and intelligent decisions and case management to provide processes with enough flexibility for end-to-end case management success. Another key feature of IA is the integration between the different systems used in the company. Integration prevents the duplication of data in the systems, and users will need to work on only one platform. RPA follows rules to automate work that has no variation and is great for repetitive tasks that often string together to form a simple business process. Intelligent Automation integrates all the capabilities of RPA and capabilities of BPM and is used to:

- Process unstructured data, automate tasks that require judgment, and detect and adapt to constant change.
- Help plan resources, set achievable KPIs and help ensure optimal operational results.

- Handle exceptions and reduce bot management.

4.1 Intelligent Automation Tools

The Shared Services organizations should be armed with a variety of automation tools. This helps increase flexibility and the ability to handle changing operational needs and customer expectations for service levels and cost management. Today's automation tools have been used successfully in the private sector and are now starting to be deployed broadly within government agencies. The spectrum of tools, which includes Robotic Process Automation, Cognitive Automation and Artificial Intelligence, is becoming increasingly intuitive, skilled, and sustainable as vendor competition drives improvements (Clarke, Digital automation tools: Helping improve capacity and efficiency in shared services, 2022). A description of these automation tools is as follows:

- **Robotic Process Automation** is a commercial off-the-shelf software that automates human activities for repetitive, rule-based tasks. The software typically operates in the user interface layer, where it cannot compromise underlying information technology infrastructure, systems, and business rules. These automated processes precisely follow prescribed protocols and procedures, enabling decreased human error, improved accuracy, and increased cost savings. For example, one government agency is deploying process robotics to automate the input of budgetary and continuing resolution information into their financial system as well as to track funding levels by office. This automation has freed agency staff to perform budgetary analysis instead of executing routine data entry. The result is a more productive team completing more mission-critical work at a faster pace.
- **Artificial Intelligence** is the apex of digital automation tools. AI mimics human intelligence to help solve complex problems and takes cognitive a step further by making predictions, presenting recommended courses of action based on its own analysis, and acting to implement a solution. AI could enable SSOs and their customers to address their most challenging problems, such as synthesizing extremely complex and multifaceted data sets, automatically implementing process improvements, or making data migration recommendations for potential customers.

4.2 Why Shared Services Need RPA and Artificial Intelligence?

Knowing the right mix of technology to solve business problems is key to building a successful shared services strategy. Shared Services leaders should understand the fundamentals of Robotic Process Automation (RPA) and Artificial Intelligence (AI) technologies and how these technologies can work together to get the best outcomes for their respective organizations.

Companies that have already been investigating RPA over the past few years will be aware that something new is on the horizon, vying for attention: Artificial Intelligence, or so-called Cognitive Computing, still yet relatively unknown and even less implemented, is nonetheless pushing its way onto service delivery agendas as a means of, if not leapfrogging, certainly scaling, and leveraging RPA's base capability.

Nevertheless, the combined opportunities presented by Cognitive, Machine Learning, and Artificial Intelligence herald an even bigger bang than robotics, given their possible impact on integrated decision-making.

4.3 Intelligent Automation

Intelligent automation combines artificial intelligence, including natural language processing, machine learning and decision networks, with robotic process automation. As technologies such as voice recognition, natural language processing and machine learning have become more sophisticated, the range of business issues to which intelligent automation can be applied has expanded. As a result of rapid advances in Intelligent Automation capabilities and at a lower cost, there are many potential applications for IA within financial services organizations ranging from automation of decisions (fraud, underwriting, claims, anti-money laundering, credit approvals) to automated classification and processing of documents (contract management, income verification, loan processing, regulation horizon scanning) and to voice recognition. Figure 4.1 shows the relationship between intelligent automation and technologies.

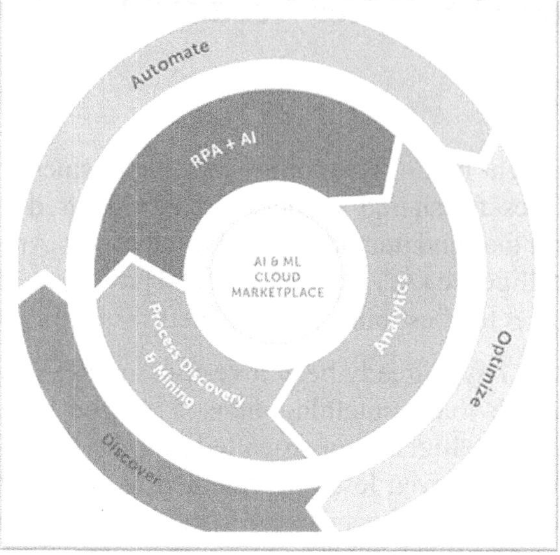

Figure 4.1: Relationship of Intelligent Automation with Technologies

Automating purely to reduce cost is not transformation; it is optimization. Companies should be looking beyond cost reduction and envisioning long-term success. IA spans the entire automation journey, from discovery, automation, and optimization to automating all front-office and back-office business processes. To extend the horizons of business process automation by order of magnitude, IA combines the task execution of RPA with Machine Learning technology (Automation, 2020). The IA cycle begins with process discovery, where AI-powered tools automatically observe people's work activities, identify optimal workflows, and propose an automation path. At the automation stage, intelligent automation applies AI to access unstructured information, including human chat conversation, audio, and video, which is crucial to making end-to-end automation possible.

The main objective of Intelligent Automation is to improve the customer and employee experience and boost productivity. In addition, it generates time and cost savings, significantly reducing human intervention in the process cycle, which allows employees to have more time to focus on creative improvements, strategy, making decisions, etc. Another objective is the reduction of errors in the processes, as well as a reduction in paper usage, as it is completely oriented towards purely digital management.

Figure 4.2 shows the development of capability along the Intelligent Automation spectrum.

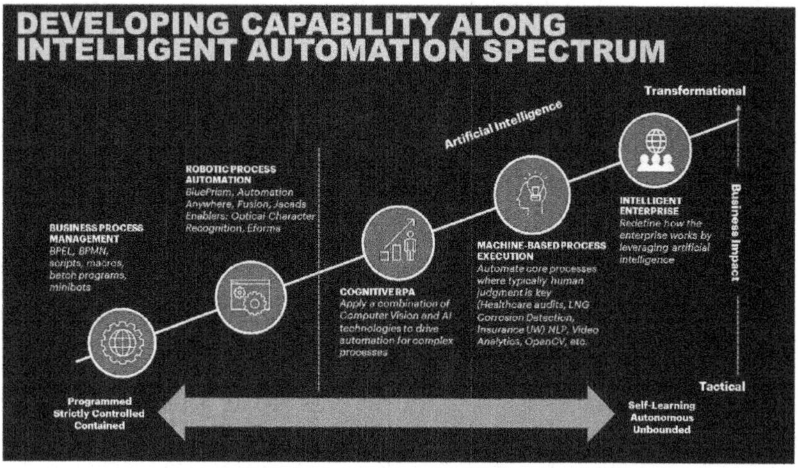

Figure 4.2: Intelligent Automation Spectrum (Accenture)

The convergence of technologies produces automation capabilities that elevate business value and competitive advantage for organizations.

4.4 Benefits of Intelligent Automation

Intelligent Automation combines the capabilities of RPA and AI, which allows it to deliver value to companies faster and provide additional benefits compared to traditional automation approaches. The following are several benefits of automation that can be seen in many transformation projects:

- **Cost Reduction**. Significantly bringing down the cost of business support functions across a company is usually seen as the top benefit of automation. This also is often the easiest one to measure.
- **Higher Accuracy**. Usually, automated tasks involve processing a lot of data, which is challenging and prone to errors.
- **Increased Focus on Core Competencies**. Automating tedious, time-consuming manual tasks allows employees to focus more on their core competencies. This also leads to considerably higher customer satisfaction ratings.

- **Improved Productivity**. Minimizing human errors means less time spent making corrections, which leads to higher productivity. Also, the workforce of software bots can be scaled up or down very quickly.
- **Better Compliance**. Actions performed by the software bots are easier to audit, which makes it easier to track and correct shortcomings in processes.

Intelligent Automation reduces the burden of repetitive tasks and empowers humans with smart technologies and agile processes for quicker and more intelligent decisions. Other benefits of IA can be summarized as below:

- It increases process efficiency with a reduction in the human intervention required for repetitive tasks.
- Automation can lead to quicker response times and improved customer experience and satisfaction.
- Back-office operations can be optimized with faster data analysis and decision-making. Automated work would mean that there won't be any breaks, and the system can be running 24*7.
- With lesser human involvement, monitoring the processes and tasks becomes more efficient, and the chances of any fraud are also reduced.
- It lessens the chances of human error, thus reducing the risks and costs of the human workforce, which would otherwise be incurred.
- It will reduce the costs required to train human resources, thereby saving the costs of training and time both.

  The benefits of IA solutions also include more effective monitoring and fraud detection, product and service innovation, competitive advantage, increased revenue, and work force productivity.

4.5 Intelligent Automation Implementation

Intelligent Automation is a comprehensive approach to automation with a focus on not just mimicking human action but it includes simulating human decision-making. Intelligent automation also has the potential to embrace areas where machines can do even better than humans. The new possibilities created by intelligent automation mean work should be redefined by the outputs and problems the workforce solves, not the activities and tasks executed. An organizational structure conducive to the implementation of intelligent automation can be a combination of a central CoE (Center of Excellence) team and a federated BU (Business Unit) implementation team, especially for bot implementation. The RPA developers can be embedded in the process team at the BU level. This structure, where automation efforts are federated with the central CoE team, aids in a collaborative working culture. Intelligent Automation consists of the implementation of the following technologies:

1. **Robotic Process Automation (RPA)**

   Robotic Process Automation is a technology that aims to reduce human intervention in computer applications, especially in repetitive tasks that vary very little in each iteration. RPA works primarily by interacting with "high-level" applications, which are the software layers at the graphic interface level, as opposed to machine language or programming code. Put more simply; it is a type of software that emulates the real interaction that a human would have with conventional computer applications.

2. **Artificial Intelligence (AI)**

   Artificial Intelligence is the simulation of human intelligence by machines. In other words, it is the discipline that tries to create systems capable of learning and reasoning like a human being. Artificial Intelligence encompasses other concepts such as Machine Learning, Deep Learning, Natural Language Processing (NLP), Visual Recognition, Big Data, etc.

By using a combination of AI and RPA, companies can automate end-to-end processes. A typical end-to-end process involves both structured and unstructured data, e.g., taking in the document using AI, classifying, understanding meaning, and passing on the required action to RPA. Intelligent Automation represents the next generation of enterprise automation solutions. It is the bundling of capabilities within an automation toolbox to extend the scope and degree to which processes can be optimized while also maintaining the natural balance between work allocated to machines and people to maximize productivity. Furthermore, Intelligent Automation can be understood and designed by business users, positioning the enterprise not only to drive productivity with a collaborative machine/people workforce but also to operate automation technologies in a transparent manner. To realize the return on investment (ROI), business users must strategically identify how and where to apply Intelligent Automation and the appropriate capabilities to leverage. During early/exploratory stages, thinking about processes via the lens of a simple Acquire-Process-Analyze-Deliver (APAD) framework can initially inform what automation capabilities to focus on while maintaining an appreciation for the business problem at hand (Kofax, 2019): Acquire: If a process begins with input data, how is that data ingested? Is it typically scanned in? Is it received in digital format via email? Is it in a standard format? Or is there variability? Asking these questions can illuminate whether it makes sense to use a capability like RPA to handle data ingest versus cognitive capture or even a virtual voice agent.

- **Process:** Once data has been ingested, what needs to happen with it? Does form entry take place across multiple systems? Does it need to be moved into a specific database? Answers to these questions will suggest the extent to which BPM workflow capabilities or RPA make sense within the Intelligent Automation solution.
- **Analyze:** After the initial processing of data, or even during processing, is there a requirement to synthesize information from many sources to draw an insight or make a prediction? Is it necessary to measure the sentiment or tone within processed data to influence what happens downstream? Clarity here will inform the degree to which a machine learning and advanced analytics capability is needed.
- **Deliver:** How should information be delivered back to the business user or the customer? What information points are most important to them, and what action do they need to take? This will inform not only what is delivered back to the appropriate personnel but also *how an* activity that takes place within an automated process is reported to drive transparency in automated processes and foster greater integration between automation and people.

Intelligent Automation is radically changing the future of work. Previous periods of the Industrial Revolution focused on humans and physical machines working together, coupled with technological advances like steam power and electricity. The next revolution is marked by enterprises that can intelligently and successfully deploy multifaceted automation solutions that work closely with their people to improve how services are delivered, how ideas are generated and how customers are served. Whether Intelligent Automation is leveraged directly to improve service levels to customers or more internally to recapture time for employees to spend focused on more mission-oriented work, the results can propel the enterprise ahead and strengthen its competitive advantage.

As more and more business processes become automated, best practices dictate that IT remains consulted and informed to ensure that the automation solutions work securely, consistently and in a scalable manner. For instance, IT needs to assume responsibility for hosting enterprise automation technologies, ensuring enough memory and processing power to run them, and integrating automation within the security framework. In other models, business stakeholders can be empowered to build automation solutions within a framework established by IT. These business users design, build and monitor the performance of bots themselves, working closely with IT, who then ultimately deploy the bots. This approach is better as it allows the business to take ownership of RPA and work in a way that best fits their needs. The integration of different AI features with RPA helps organizations extend automation to more processes, making the most of not only structured data but especially the growing volumes of unstructured information. Unstructured information, such as customer interactions, can be easily analyzed, processed, and structured into data useful for the next steps of the process, such as predictive analytics.

Cost reduction and workforce and process efficiency are a few benefits driving major interest in automation across many organizations. While most organizations are still in the early phase of RPA and AI-enabled automation adoption, developing an overarching automation strategy and developing the right use cases that map to specific business outcomes will be crucial to the successful adoption of IA.

4.6 Examples of Intelligent Automation

Intelligent Automation is applicable in a wide variety of applications. The application examples of IA can be divided into Decision Making and Physical Tasks in two categories, as follows:

- **Financial Services:** Investment managers use IA to go through research data to identify consistencies. IA systems are capable of spotting inconsistencies through a vast volume of data that would have been otherwise impossible for a human to go through.
- **Health Care:** With the enormous amount of data available in medical research, robots could suggest alternate treatments based on the symptoms and medical history of the patient. This could lead to faster diagnosis and treatment of the patient, as opposed to a doctor who would have to invest many hours to investigate the case. The final decision regarding treatment would ultimately rest with the doctor. WellPoint, a health insurance provider in the US, has implemented a utilization review system that utilizes IBM's Watson cognitive computing technology. After about 15000 hours of training, the system is now able to understand the treatment prescriptions in unstructured English and match them to WellPoint's medical policies and guidelines. This makes the system able to provide quicker and better-informed decisions regarding the patient's care and saves significant time and costs.
- **Workflow Software and Conditional Logic:** There are always times in any process, even though automated when the consecutive steps depend on the logical outcome of the previously completed tasks. A typical automated system might not be able to tackle such tasks. Intelligent automation systems can analyze the pre-existing data regarding similar situations in the past and decide the direction of the workflow.
- **Distributing Products:** Kiva Systems developed robots for companies such as Crate & Barrel and Walgreens that can work and travel in warehouses without colliding with any obstacle, a fellow robot or a human while working on the floor. These robots can think for themselves and improve the efficiency with which they are able to fulfill orders.

- **Robots Collaborating with Humans:** Robots working on the assembly line aren't a new concept. However, earlier, robots were developed and programmed to complete specific tasks and were thus kept segregated from human workers for safety reasons. Volkswagen has introduced collaborative robots that can recognize human workers nearby and react to the situation accordingly. Thus, humans and robots are now capable of working together in harmony without concern for safety.
- **Driverless Cars:** Autonomous vehicles are in the news for all the right reasons. Driverless cars improve the safety on roads better than human drivers as they reduce the possibility of human error or distraction during driving. Autonomous driving technology by Google has empowered numerous cars already.
- **Strategic Processes**: Sales, underwriting and claims decisions for large and complex risks, risk inspection, and loss verification, which are intellectually more demanding, are best done by specialist underwriters and claim adjusters.
- **Knowledge-based Processes** that require some intelligence, like underwriting & claims decisions for medium complexity cases, policy change requests involving financials, renewal processing, and coverage query resolution, can be partially automated, with limited intervention from generalist underwriters and claims adjusters.
- **Transactional Processes** like data entry of risk and claim information, automated underwriting and claims settlement for simple, standard cases, communication letter generation, etc., have a very high potential for automation.

Intelligent Automation represents an evolved version of automation in which machines mimic human actions and possess cognitive capabilities, including natural language processing, speech recognition, computer vision technology, and machine learning. Such machines with automated intelligence comprehend the vast amount of structured and unstructured data, analyze, understand, and learn it on the go, and intelligently automate processes to bring in more operational efficiency as well as business efficiency. The concept of automation in the digital world is evolving, and technology is progressing, adding more capabilities of human brains into machines day by day. Intelligent Automation technology differs from normal IT automation, given the added layer of Machine Learning and

Artificial Intelligence capabilities integrated with it. The self-learning and application of intelligence bring in much more efficiency in business with respect to time and effort and elevate overall customer experience.

4.7 Challenges of Intelligent Automation

Intelligent Automation and RPA tools automate repetitive and mundane tasks, freeing up employees to do more high-value work. In turn, this helps organizations deliver better customer experiences, increase business agility, and improve productivity. Many companies are struggling to realize these benefits due to technology's real and perceived challenges. Although IA and RPA have applications in almost every industry, many of the logistical and technological challenges businesses face, from making the business case to automating to implementing and embedding the technology, are poorly understood. These innovations are driving digital transformation but there are also challenges:

- **Need for Data Governance:** IA requires a lot of data to perform smart operations. Companies need to focus on data as an asset. Data governance, which is a mix of rules and tools designed to improve integrity, availability and data readiness, is a challenge.
- **Need to Hire Data Scientist:** IA uses a different kind of software that mimics the human thinking process and decides in a probabilistic way. As a result, there is a need to hire a data scientist than a programmer.
- **Need for Reskilling the Workforce:** IA requires a transformation of the workforce, which means that there is a need to reskill the jobs.
- **Need for Scaling the Technology:** IA has individual sensory capabilities, from using image recognition to scan photographs to converting the spoken word to text to predicting the future based on past actions. However, scaling this technology is essential to achieve true business transformation.
- **Security Issue:** For IA, the biggest security issue arises at the point where humans and machines interact. For example, human error during an automated financial reporting process can result in a loss of weeks and a delay in corporate reporting. Other security issues that need to be considered include rogue access, data loss, hacking, privilege abuse, vulnerabilities, and malware. There is a need to use

security protocols like encrypting data and multiple layers of authentication, along with reducing access rights and requiring human validation on certain processes.

4.8 Conclusion

Shared Services organizations can use Intelligent Automation tools to integrate disparate and disconnected systems, automate processes and transform operations as they continue to grow in complexity and cost. These organizations have played a significant role in leading the charge to source and implement Intelligent Automation solutions. For many, the scope of projects where Shared Services organizations have played a significant role also includes projects that are outside the scope of Shared Service processes. The implementation of Intelligent Automation solutions is an excellent example of Shared Services being a true partner in the company, providing leadership where needed based on a unique appreciation of the opportunities at play.

Intelligent Automation offers a more comprehensive and real-time view of business performance to the organization's benefit, brings new levels of measurability, supports compliance with automated processing and reporting and enables functions to focus on delivering a better customer experience for internal and external audiences. Moreover, the iterative and granular nature of implementing IA tools means a modest upfront investment can deliver a significant return.

The transformative benefits of intelligent automation can be achieved to their fullest extent when technology and tools are matched by deep understanding and experience not only of their implementation but also of the processes to which they apply inside global operations. Intelligent automation is a term related to digital transformation with the advantage of being better defined and providing solutions by combining technologies such as BPM, RPA, and artificial intelligence. An ideal case would be a company with a backbone orchestrated by workflows (using BPM) that involve users, systems, data, and documents; uses RPA in specific moments and tasks (to tackle bottlenecks in previously manual tasks); with hybrid decision-making between managers and automation based on Artificial Intelligence.

The main objectives of Intelligent Automation are to improve end-to-end customer experience, improve the productivity of employees, save time and costs, get rid of repetitive tasks and enable employees to focus on tasks that add value for the company, such as creative improvements, new strategies and solutions and decision making.

CHAPTER 5: ROBOTIC PROCESS AUTOMATION

Robotic Process Automation (RPA) is the concept of using a software platform of "virtual robots" to manipulate existing application software in the same way that a human does to process a transaction. Imagine a world in which the meaning of "work" has been redefined for millions of people, where our service economy can focus on providing services delivered by an engaged talent pool that innovates such services. In this new world, work would no longer be associated with functioning within repeatable systems with mundane transactional processes. Instead, this other world would have workers who re-think end-to-end processes on a more holistic level with the goal of simultaneously impacting several factors: quality, efficiency, cost-effectiveness, functionality, customer satisfaction and compliance while creating remarkable delivery of business support services.

Back offices are where the operational support systems for services are created, managed, and delivered. Back offices are always under pressure to contain costs in highly competitive industries like insurance and financial services, but cost efficiency must be balanced with other performance imperatives such as service excellence, business enablement, scalability, flexibility, security, and compliance. Using years of research on back offices, it has been learned that low-performing back offices can be transformed into high-performing back offices through six transformation levers: centralized physical facilities and budgets, standardized processes across business units, optimized processes to reduce errors and waste, work relocation from high cost to low-cost destinations, technology enablement with, for example, self-service portals, and automated services (Willcocks, Leslie P.; Lacity, Mary C., 2016).

RPA is relatively low cost, quick to implement and unobtrusive; thus, it starts what will likely be one of the important technologies in the future regarding how automation will change the value proposition in all organizations (Gartner, 2016).

5.1 What is Robotic Process Automation?

Major business functions that offer opportunities for process improvement through RPA include Purchasing and Supply Chain Management, Sales, Finance and Accounting, IT and Human Resources. RPA is ideal for preventing error rates, reducing variability, improving cycle time, and increasing productivity in processes that follow a standard procedure with minimal deviation. Software Bots can perform repetitive, monotonous, high-volume tasks, freeing workers to focus on activities that require higher-order thinking. High-volume bulk processing functions within tools such as enterprise resource planning (ERP) systems or core databases are ideal candidates for RPA, as are desktop applications and workflows that require information gathering from multiple sources. For tasks too small and diverse for IT changes, RPA offers an alternative to outsourcing and offshoring (Smith, 2015).

Finance leaders obsess about transformation levers such as people, processes, and technology. However, today's solutions save the requisite investment in an expensive ERP platform and have typically leaned heavily on the equation of people and process. Less expensive resources located offshore delivering services, plus added process improvement, equals significant savings. Up until now, the technology component has been limited to communication widgets that facilitate workflow and e-invoicing.

RPA is the application of flexible tools to automate manual activities for the delivery of business services, resulting in improved end-to-end process efficiency and effectiveness. RPA is most applicable to rules-driven, data-intensive processes that are repetitive in nature. The application of RPA can cross multiple systems and include multiple decision points/calculations. RPA applications require an electronic input or trigger to commence working, yet the underlying technologies are still emerging, each taking a different approach (ACCA, 2015).

Transactional processes are typically located within a Shared Services center or another part of the back office. RPA does not replace Business Process Management (BPM) but rather complements it. RPA and BPM are each suited to automating different types of processes. BPM solutions are best suited for processes requiring IT expertise on high-valued IT investments like ERP and Customer Relationship Management (CRM) systems. The distinguishing attributes of RPA software are that it is designed for non-programmers to use and it does not disturb existing systems. This means that the threshold of business processes worth automating is substantially lowered. RPA solutions are typically deployed by business operations staff with IT oversight (but not with IT developers) for processes that require business and process expertise. The significantly lower IT investment costs now make automating these processes financially beneficial. Pat Geary, CMO for Blue Prism, said: “We are not trying to replace enterprise IT, and we are not really trying to compete with BPMS. It’s really this long tail of processes that are typically deployed by humans that are most suitable for RPA. Humans can be redeployed to more intelligent decision-making tasks” (Willcocks, Leslie P.; Lacity, Mary C., 2015).

The evolution of digital automation includes a spectrum of RPA, intelligent automation, and cognitive automation. For many processes, RPA can take the human touch point out of the equation and eliminate subjective errors. The RPA experts conclude that rather than cutting back on Full Time Equivalents (FTEs), staff can be redeployed into roles where free of the routine transactional work, they can leverage their know-how and expertise to deliver far greater value. While RPA can drive down errors, duplications, and high processing costs, the additional benefit of refocusing staff on analytics in Shared Services organizations will result in the availability of increased business insights and intelligence that can be leveraged for improved decision-making. RPA, therefore, provides an alternative higher-level career path that aids in the retention of an organization’s best talent by offering the opportunity to develop new, in-demand skill sets.

5.2 Software Bots

A Software Bot is a fundamental element of RPA technology that can be deployed to automate repeatable, mundane, rules-based workflow process tasks across multiple functions in an organization, including Shared Services. RPA may sound misleading because no physical robots wandering around offices are involved in performing human tasks. Rather, RPA is a software solution that can reside either on a server or in the cloud. RPA software programs are, in some cases, faster and more accurate than what humans can achieve. Software Bots have super-human stamina, offering 24/7 productivity with no processing handoffs due to the shift changes.

A Software Bot or "Bot" captures and interprets existing applications to enable transaction processing, data manipulation and communication across multiple IT systems. Multiple Software Bots can be seen as a virtual workforce, working as a back-office processing center but without the human resources. Software Bots undertake processes like human counterparts and can work on multiple processes just like a Shared Services staff member can learn to work on accounts payable and travel and expenses processes. The robots use a 'virtual machine' and dedicated logins to interact with different applications and systems in the same way as human teams (Horton, 2015).

Software Bots can be scaled up quickly to seasonal or unanticipated spikes in work volume. They are fundamentally reshaping the world of Information Technology Outsourcing (ITO) and Business Process Outsourcing (BPO) and now the time seems right for adoption by Shared Services organizations. Process standardization, improvement in service quality, and cost reduction are the most important outcomes of automating processes in shared services. Organizations still feel that if they were to further automate their processes, they would use ERP and/or other enterprise management systems instead of Software Bots. They think that RPA is not the solution to all needs as ERP and BPM can be alternate ways to automate processes.

Examples of functional processes in Shared Services, which are good candidates for automation using Software Bots, are shown in Figure 5.1.

Finance Function	HR Function	IT Function
Accounts Payable	Payroll	Ticket Management
Order-to-Cash	Hire-to-Retire	Database Management
Procure-to-Pay		
General Ledger		
Financial Analysis		
Financial Reporting		

Figure 5.1: Candidates for Automation Using Software Bots

In summary, the key benefits associated with Software Bots include increased speed of operations, elimination of errors, utilization of FTEs on more value-added activities, increased customer satisfaction, reduction in FTEs, improved customer service, increased quality, and increased agility to handle more work during peak periods.

5.3 Benefits of Implementing RPA

RPA has driven substantial reduction in the cost of services, primarily through automating repetitive tasks currently performed by humans. RPA offers companies benefits that include fast ROI, flexibility, and security. Potential applications of RPA solutions include traditionally labor-intensive areas such as service desks and customer care centers, as well as a broader range of functions in IT operations, such as network, storage, server and application management, database administration, virtual machine provisioning, process orchestration and teleconferencing. Service delivery model options include partnering directly with a software provider that licenses smart software or with a provider that offers a home-grown solution (Fausti, 2015).

Robotics tools are fast, easy, and inexpensive to implement, with development in six to eight weeks, which is typical for faster realization of return on investment. Outsourcing consultancy firm Everest Group reports that RPA can reduce costs by up to 65%. And RPA's ability to log data at the transactional level enables ongoing decision-making that is fast, accurate and predictive. Depending on project requirements, robotic tools can be developed either in "batch mode" to complete end-to-end processes or, when human intervention is indispensable, in "assisted mode." RPA can be integrated with multiple applications at the presentation layer, ensuring that the Software Bot does not modify or enhance clients' applications. It also carries no risk of unauthorized data access; since the business function leverages the already-available underlying application, access authorization concepts are automatically inherited.

RPA is an evolution of technology in support of automating processes. One of the enormous potential advantages of RPA is that it offers an affordable, easy-to-implement and cost-effective solution. The benefits of RPA are shown in Figure 5.2.

Figure 5.2: Benefits of RPA

These benefits include increased speed of operations, elimination of human errors, utilization of FTEs on more value-added activities, increased customer satisfaction, reduction in FTEs, improved customer service, increased quality, and increased agility to manage more work during peak periods. A detailed description of all the benefits offered by RPA is shown below:

- **Reduced Human Involvement:** RPA can help the organization manage its time and prioritize more important tasks.
- **No Interruption of Work:** Bots never get tired, so tasks can be carried out all day and all night.
- **Optimal Utilization of Human Resources:** RPA implementation helps to optimize the way employees cooperate with each other and manage their time. More importantly, RPA saves the organization from recruiting just for growth because bots will handle most routine tasks.
- **Preciseness:** Close to zero error rates make a huge difference when financial operations are manual.
- **Operation Speed:** Compared to humans, bots are much faster at executing program code.
- **System Agnosticism:** RPA bots do not rely on the specifics of the operating system but on the application. The bots are aimed at functioning in the UI, being a cross-platform app, or working from a cloud.
- **Cost Effectiveness:** Operating costs are lower with an army of virtual workers. But, more importantly, the return on investment (ROI) is high.
- **Data Quality Improvement:** Performing business processes and employees source data from various systems. Making errors or structuring them in different ways impacts the quality of data, which makes further analysis more difficult. A bot is not prone to making mistakes.
- **Fast Implementation:** In most cases, the bots do not require any integration with the existing systems like ERP or CRM.
- **Scalability:** Bots can be multiplied or deleted, which allows for scaling the workforce as required, making RPA suitable for both small and large businesses.
- **Ease of Use:** Some of the vendors offer free bots that can be configured and installed like any other software program and require no code to configure it.

- **Wide Sphere of Usage:** Because RPA bots read instructions, creating bots with an industry-dependent standard pack of default routine tasks is possible. That makes RPA universal if it can be used to automate nearly any routine process.

It's obvious that the advantages of RPA override the limitations. However, reliance on human interaction is still an issue, a problem which can probably be solved with the help of Artificial Intelligence. To maximize benefits from the implementation of RPA, an organization should start where there is volume and then centralize, standardize repeatable processes, run a pilot, optimize for immediate value, and optimize for future value. A responsibility relationship chart should be developed to clarify the roles and responsibilities of business units and IT during the implementation of RPA so that the process owners own the business processes and technical issues are addressed by IT.

5.4 Challenges Facing Robotic Process Automation

RPA implementation in an organization is a very special and crucial task, and it requires excellent coordination between all levels of the employees involved in process automation. It needs strong leadership who can lead the team to the completion of the automation process. Coordination between the teams not only helps in completing the task with the utmost accuracy but also helps the company create a system that works with RPA rather than against it. The involvement of IT teams in building the architecture will minimize the risk factor and ensure the security levels. While implementing Robotic Process Automation, there are a few challenges that every company faces and needs to overcome. The list of top challenges is as follows:

1. **Unclear Division of Responsibilities:** Robotic Process Automation involves a lot of responsibilities, such as responsibilities between IT, functions, and business units. Also, there are process ownership responsibilities for every process that is being automated.
2. **Employee Resistance:** The challenge of employee resistance is real, as employees need new skills, frequent communication, and a clear understanding of the processes.
3. **Infrastructure Issues:** One of the top challenges of RPA implementation is the need for the infrastructure to be implemented and deployed at the desired location.
4. **Business Case Issues:** Deployment and implementation of Robotic Process Automation affects processes at many levels. The processes cut across organizational boundaries, and the assumptions made in the business case might not be addressed depending on the companies on different levels. As the implementation of RPA solutions requires skills with great effort and dedicated training to automate processes, the economic effect should be considered.
5. **Unrealistic Expectations:** The enthusiasm associated with RPA can cause business leaders to have the impractical belief that just about everything can be successfully automated. To counter this thinking, a best practice is to be conservative when clarifying success criteria and be realistic in defining the business outcomes. There is a need to focus on the high-volume, low-complexity tasks where bots can make an immediate difference.
6. **Ownership:** Ownership refers to who owns RPA solutions. Is it led by IT or Business teams? Business needs to provide the requirement, approve the solution design for feasibility, help in UAT and then measure the success rate. IT teams have a limited role, primarily restricted to providing support in infrastructure requirements and test data creation. Ownership also consists of all the responsibilities; for example, the decision-making process needs to be clearer and sharper, every team must have a leader who has the required knowledge to understand all the processes, etc.
7. **Limited Application:** There are documents such as handwritten documents. Partial automation based on cost and benefit analysis can

be considered if full end-to-end automation is not available or reachable within a reasonable timescale.

8. **Short-term Solution**: Robotic Process Automation provides immediate relief and benefit, but in the long term, it requires complex platforms and frequent workflow review and optimization.

After the implementation has been done and all the processes have successfully been automated, RPA challenges or difficulties faced during operations include:

- It is easy to automate simple business processes. However, complex business processes need a greater understanding of process complexity, followed by testing and rework cycles. Lack of real-time visibility of the process in which the software robot works will lead to failure. Most of the time, failures of RPA adoption are due to the gap between expected results and the actual functioning of software robots.
- Processes that are prone to frequent change may not qualify for automation, given the change management cost and time to manage the change. This makes it worse if the process is automated based on the usage of internal sources of applications or systems.
- It is not always simple to understand business workflow and processes. Lack of operational knowledge and lack of real-time visibility on the part of the process the robot is working- RPA tools work best on a virtualized desktop environment, with appropriate scaling and business continuity setup. Another common mistake is targeting RPA as a highly complex process, which results in significant automation costs.
- Automating an "as is" process without checking the viability of the process in the future may not yield the desired results.
- To maintain the output, RPA solutions require maintenance for rule-based or repetitive tasks.

5.5 Conclusion

RPA is a digital workforce of software robots that can execute any business process. In contrast to other traditional IT solutions, RPA allows organizations to automate at a fraction of the cost and time previously encountered. RPA is also non-intrusive in nature and leverages the existing infrastructure without causing disruption to underlying systems, which would be difficult and costly to replace. RPA tools/vendors are the software through which tasks can be configured to get automated. Unlike other forms of automation, RPA has the intelligence to decide if a process should occur. It can analyze data presented to it and decide based on the developer's logic parameters. In comparison to other forms of automation, it does not require system integration.

Technological advancements are happening in every field, and conducting business has become easier than ever before. The invention of robots is one of the important landmarks in the history of technology. Robotic Process Automation is shifting humankind to perform more efficient and skilled work rather than mundane and repetitive tasks. RPA will enhance the business process by eliminating the hindrances that are revealed from the regular conduct of business. RPA can be an excellent tool for instant reduction of business costs and improving efficiency. Being aware of challenges that are associated with RPA implementation and having a perfect plan for overcoming them would result in the success of RPA. The key challenges faced when organizations implement RPA are unclear division of responsibilities between IT and functional organizations, lack of understanding of what RPA means and where it can be applied, lack of management support and weak change management process across the organization. Other challenges faced when implementing RPA include properly dealing with business disruption and employee anxiety.

Robotic Process Automation can be a fast, low-risk starting point for automating processes that rely on outdated legacy systems. Bots can pull data from manual systems without APIs into digital processes, ensuring faster and more efficient outcomes. RPA does not transform the organization by itself and is not a fix for enterprise-wide broken processes and systems. Used the right way, RPA can be a useful tool in the digital transformation toolkit. Instead of wasting time on repetitive tasks, people are freed to focus on customers who get the desired outcomes more quickly. RPA offers the opportunity to automate repetitive processes and achieve business goals more efficiently. Deploying software robots can enable organizations to execute current processes faster and more accurately, 24/7. The benefits can be significant and include boosting the return on investment, improving operational accuracy, enhancing compliance, achieving cost reductions, and increasing scalability.

CHAPTER 6: ARTIFICIAL INTELLIGENCE

In computer science, Artificial Intelligence (AI), sometimes called machine intelligence, is intelligence demonstrated by machines, in contrast to the natural intelligence displayed by humans and animals. Artificial Intelligence represents the field of computer science that strives to create ever-smarter machines that can solve problems thought once only solvable by humans. More precisely, it seeks to create machines that can generate a higher statistical probability of a correct output faster and more reliably than a human. AI allows machines to learn from experience, adjust to new inputs, and perform human-like tasks.

6.1 Artificial Intelligence Offerings

ChatGPT or similar Artificial Intelligence (AI) offerings are moving the AI adoption needle. ChatGPT is an AI chatbot developed by OpenAI and released in November 2022. The name "ChatGPT" combines "Chat," referring to its chatbot functionality, and "GPT," which stands for Generative Pre-trained Transformer, a type of large language model. As companies incorporate AI into their business operations, spending on AI-centric systems will rise at a compound annual growth rate of 27% between 2022 and 2026 to exceed a staggering $300 billion (Clark, 2023).

6G connectivity is waiting in the wings to help companies make AI models more efficient and responsible while data from machines and connected devices continues to grow. Digital transformation has resulted in a shift across all industries towards Robotic Process Automation (RPA), Artificial Intelligence (AI), and utilizing Intelligent Automation (IA) to streamline business operations. Leveraging AI and the benefits of IA has freed employees from low-value and time-consuming tasks. To remain competitive in the technology-driven business world, it is critical that businesses make evaluating their automation journeys a top priority. AI refers to the simulation of human intelligence in machines programmed to think like humans and mimic their actions. The term may also be applied to any machine that exhibits traits associated with a human mind, such as learning and problem-solving.

Artificial Intelligence systems typically demonstrate at least some of the following behaviors associated with human intelligence: planning, learning, reasoning, problem solving, knowledge representation, perception, motion, manipulation, and, to a lesser extent, social intelligence and creativity. So, AI is defined as the theory and development of intelligent computer systems able to perform tasks that normally require human intelligence (i.e., visual perception, speech recognition, decision-making, translation between languages). AI is the simulation of human intelligence processes by machines, especially computer systems. These processes include learning (acquiring information and rules for using the information), reasoning (using rules to reach approximate or definite conclusions), and self-correction.

AI is ubiquitous and is used to recommend what one should buy next online, to understand what is said to virtual assistants such as Amazon's Alexa and Apple's Siri, to recognize who and what is in a photo, to spot spam, or to detect credit card fraud. At an extremely high level, AI can be split into two broad types: narrow AI and general AI. Narrow AI is what is seen all around in computers today: intelligent systems that have been taught how to conduct specific tasks without being explicitly programmed how to do so. This type of machine intelligence is evident in the speech and language recognition of Siri's virtual assistant on the Apple iPhone, in the vision-recognition systems on self-driving cars, and in the recommendation engines that suggest products based on what was bought in the past. Unlike humans, these systems can only learn or be taught how to do specific tasks, which is why they are called narrow AI.

There are a vast number of emerging applications for narrow AI: interpreting video feeds from drones carrying out visual inspections of infrastructure such as oil pipelines, organizing personal and business calendars, responding to simple customer-service queries, coordinating with other intelligent systems to carry out tasks like booking a hotel at a suitable time and location, helping radiologists to spot potential tumors in X-rays, flagging inappropriate content online, detecting wear and tear in elevators from data gathered by IoT devices, the list goes on and on.

General AI is quite different and is the type of adaptable intellect found in humans. It is a flexible form of intelligence capable of learning how to conduct vastly different tasks, anything from haircutting to building spreadsheets or reasoning about a wide variety of topics based on accumulated experience. There is a broad body of research in AI, much of which feeds into and complements each other (Heatth, 2018).

6.2 How Artificial Intelligence Works?

Artificial intelligence makes it possible for machines to learn from experience, adjust to new inputs, and perform human-like tasks. It combines substantial amounts of data with fast, iterative processing and intelligent algorithms, allowing the software to learn automatically from patterns or features in the data. AI is a broad field of study that includes many theories, methods, and technologies, as well as the following major subfields (Insights, 2020):

- **Machine Learning** automates analytical model building. It uses methods from neural networks, statistics, operations research, and physics to find hidden insights in data without explicitly being programmed for where to look or what to conclude.
- **A Neural Network** is a type of machine learning made up of interconnected units (like neurons) that process information by responding to external inputs and relaying information between each unit. The process requires multiple passes at the data to find connections and derive meaning from undefined data.
- **Deep Learning** uses massive neural networks with many layers of processing units, taking advantage of advances in computing power and improved training techniques to learn complex patterns in large amounts of data. Common applications include image and speech recognition.
- **Cognitive Computing** is a subfield of AI that strives for a natural, human-like interaction with machines. Using AI and cognitive computing, the goal is for a machine to simulate human processes through the ability to interpret images and speech – and then speak coherently in response.
- **Computer Vision** relies on pattern recognition and deep learning to recognize what's in a picture or video. When machines can process, analyze, and understand images, they can capture images or videos in real-time and interpret their surroundings.

- **Natural Language Processing** (NLP) is the ability of computers to analyze, understand, and generate human language, including speech. The next stage of NLP is natural language interaction, which allows humans to communicate with computers using normal, everyday language to perform tasks.

Additionally, several technologies enable and support AI:

- **Graphical processing units** are key to AI because they provide the heavy computer power required for iterative processing. Training in neural networks requires big data plus computer power.
- **The Internet of Things** generates massive amounts of data from connected devices, most of it unanalyzed. Automating models with AI will allow us to use more of it.
- **Advanced algorithms** are being developed and combined in new ways to analyze more data faster and at multiple levels. This intelligent processing is key to identifying and predicting rare events, understanding complex systems, and optimizing unique scenarios.
- **APIs, or application programming interfaces,** are portable packages of code that make it possible to add AI functionality to existing products and software packages. They can add image recognition capabilities to home security systems and Q&A capabilities that describe data, create captions and headlines, or call out interesting patterns and insights in data.

Figure 6.1 shows that Artificial Intelligence includes machine learning and deep learning, allowing machines to self-optimize performance and alert managers of potential failures or issues before they occur.

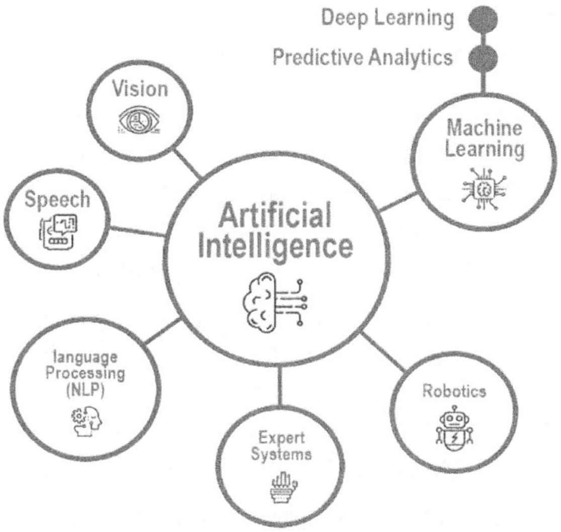

Figure 6.1: Artificial Intelligence

The use of AI will lead to better responsiveness by customer service representatives. Since the technology hinges on automation, order accuracy will increase, which will enable better spending analysis and cost reductions through more advanced analytics.

AI automates repetitive learning and discovery through data. But AI is different from hardware-driven RPA. Instead of automating manual tasks, AI performs frequent, high-volume, computerized tasks reliably and without fatigue. The goal of AI is to provide software that can reason on input and explain on output. AI provides human-like interactions with software and offers decision support for specific tasks.

6.3 Types of Artificial Intelligence

Artificial intelligence can be categorized into one of four types.

- **Reactive AI** uses algorithms to optimize outputs based on a set of inputs. Chess-playing AIs, for example, are reactive systems that optimize the best strategy to win the game. Reactive AI tends to be static, unable to learn or adapt to novel situations. Thus, it will produce the same output given identical inputs.
- **Limited memory AI** can adapt to experience or update itself based on new observations or data. Often, the amount of updating is limited (hence the name), and the length of memory is short. Autonomous vehicles, for example, can "read the road" and adapt to novel situations, even "learning" from experience.
- **Theory-of-mind AI** is fully adaptive and has an extensive ability to learn and retain past experiences. These types of AI include advanced chat-bots that could pass the Turing Test, fooling a person into believing the AI was a human being. While advanced and impressive, these AIs are not self-aware.
- **Self-aware AI**, as the name suggests, becomes sentient and aware of its own existence. Still, in the realm of science fiction, some experts believe that an AI will never become conscious or "alive."

Figure 6.2 summarizes types of Artificial Intelligence (Chazey, Creating an Appropriate Ecosystem to Advance Your Organization's Intelligent Automation Journey, 2023).

Narrow AI	Refers to AI systems designed to perform specific tasks or functions. Includes voice assistants, image recognition algorithms, and recommendation systems.
General AI	AI systems that can understand, learn, and perform intellectual tasks like a human. Can apply knowledge and skills across various domains, exhibiting human-like intelligence.
Machine Learning	The Subset of AI uses algorithms/models to enable computers to learn from data, make predictions, or take actions themselves. ML algorithms can automatically identify patterns, learn from experience, and improve their performance over time.
Natural Language Processing	Natural Language Processing focuses on enabling computers to understand, interpret, and generate human language. Encompasses tasks such as text classification, sentiment analysis, machine translation, and chatbot interactions.
Computer Vision	Computer Vision is the field of AI that enables computers to interpret and understand visual information from images or videos. It involves tasks like object detection, image recognition, facial recognition, and scene understanding.
Reinforcement Learning	It is a type of machine learning where an agent learns to make decisions or take actions in an environment to maximize a reward. It involves learning through interaction and feedback from the environment.
Generative AI	Focuses on the ability of AI systems to generate new content or data, such as images, music, text, or other creative outputs. Learns patterns from existing data and uses that knowledge to create original content that resembles the training data.

Figure 6.2: Types of Artificial Intelligence

Artificial intelligence can be divided into two different categories: weak and strong.

- **Weak artificial intelligence** embodies a system designed to carry out one job. Weak AI systems include video games such as the chess example from above and personal assistants such as Amazon's Alexa and Apple's Siri. You ask the assistant a question, and it answers it for you. Weak AI—also called Narrow AI or Artificial Narrow Intelligence (ANI)—is AI trained and focused on performing specific tasks. Weak AI drives most of the AI that surrounds us today. 'Narrow' might be a more accurate descriptor for this type of AI as it is anything but weak; it enables some very robust applications, such as Apple's Siri, Amazon's Alexa, IBM Watson, and autonomous vehicles.
- **Strong artificial intelligence** systems are systems that carry on tasks considered to be human-like. These tend to be more complex and complicated systems. They are programmed to handle situations where they may be required to solve problems without a person intervening. These systems can be found in applications like self-driving cars or hospital operating rooms. Strong AI is made up of Artificial General Intelligence and Artificial Super Intelligence. Artificial General Intelligence or General AI is a theoretical form of AI where a machine would have an intelligence equal to humans; it would have a self-aware consciousness that can solve problems, learn, and plan. Artificial Super Intelligence, also known as superintelligence, would surpass the intelligence and ability of the human brain.

Since its beginning, artificial intelligence has come under scrutiny from scientists and the public alike. One common theme is that machines will become so highly developed that humans cannot keep up and will take off on their own, redesigning themselves at an exponential rate. Another is that machines can hack into people's privacy and even be weaponized. Other arguments debate the ethics of artificial intelligence and whether intelligent systems such as robots should be treated with the same rights as humans.

Self-driving cars have been controversial as their machines tend to be designed for the lowest possible risk and the least casualties. If presented with a scenario of colliding with one person or another simultaneously, these cars would calculate the option that would cause the least damage.

Another contentious issue many people have with artificial intelligence is how it may affect human employment. With many industries looking to automate certain jobs using intelligent machinery, there is a concern that people would be pushed out of the workforce. Self-driving cars may remove the need for taxis and car-share programs, while manufacturers may easily replace human labor with machines, making people's skills obsolete.

6.4 Artificial Intelligence Models

Artificial Intelligence models are a product of algorithms and training data. While the algorithms enabling AI are complex, most underlying resources are open source. As a result, the primary differentiation between strong and weak AI models is the data that can be used to train it. This means that for any firm seeking to develop a successful AI model, securing training data is critical. Ideally, this training data would be a constantly refreshing (and growing) flow, not a 'one-time' stock of data, thus allowing the AI model to learn and develop in response to the evolving data flow.

The most competitively defensible AI models in any industry secure a unique and useful set of data and continuously draw in more training data, and in doing so, establish a scale of data that is difficult for any new competitors. The overriding need for data makes digital platform models that form a data-rich interface for developing AI models. While the fundamental dynamics of AI may be consistent across industries, it is not clear how the created pressures will reshape the structure and competitive dynamics of the financial sector, nor can it be concluded that they will have the same impact across multiple subsectors of Financial Services. Artificial Intelligence and Machine Learning provide ample possibilities for businesses to improve their operations and maximize their revenues. The array of problems the businesses face is huge, and the variety of ML models used to solve these problems is quite wide, as some algorithms are better at dealing with certain types of problems than others. Thus, one needs a clear understanding of what every type of ML model is good for, and here are the ten most popular AI algorithms (Fedak, 2018):

1. **Linear Regression** is used in mathematical statistics to find values of coefficients that provide the most impact on the precision of the function.

2. **Logistic Regression** is another popular AI algorithm that is able to provide binary results. This means that the model can both predict the outcome and specify one of the two classes of the function. The function is also based on changing the weights of the algorithms, but it differs because the non-linear logic function is used to transform the outcome.
3. **Linear Discriminant Analysis** is a branch of the logistic regression model that can be used when more than two classes exist in the output. Statistical properties of the data, like the mean value for every class separately and the total variance summed up for all classes, are calculated in this model. The predictions allow us to calculate the values for each class and determine the class with the most value.
4. **Decision Trees** is one of the oldest, most used, simplest, and most efficient ML models around. It is a classic binary tree with **a Yes** or **No** decision at each split until the model reaches the result node.
5. **Naive Bayes**' algorithm is a simple yet very strong model for solving a wide variety of complex problems. The model is called naive as it operates on the assumption that all the input data values are unrelated to each other. While this cannot occur in the real world, this simple algorithm can be applied to a multitude of normalized data flows to predict results with great accuracy.
6. **K-Nearest Neighbors** is quite a simple and very powerful ML model, using the whole training dataset as the representation field. The predictions of the outcome value are calculated by checking the whole data set for K data nodes with similar values (so-called neighbors) and using the Euclidian number (which can be easily calculated based on the value differences) to determine the resulting value.
7. **Learning Vector Quantization** is the neural network that uses the codebook vectors to define the training datasets and codify the required results. Thus, the vectors are random at first, and the learning process involves adjusting their values to maximize prediction accuracy.
8. **Support Vector Machines** algorithm is one of the most widely discussed among data scientists, as it provides powerful data classification capabilities. The so-called hyperplane is a line that separates the data input nodes with different values, and the vectors

from these points to the hyperplane can either support it (when all the data instances of the same class are on the same side of the hyperplane) or defy it (when the data point is outside the plane of its class).

9. **Random Decision Forests** are formed of decision trees, where decision trees process multiple samples of data and the results are aggregated (like collecting many samples in a bag) to find the more accurate output value.

10. **Deep Neural Networks** are among the most widely used AI and ML algorithms. There are significant improvements in deep learning-based text and speech apps, deep neural networks for machine perception and OCR, as well as using deep learning to empower reinforced learning and robotic movement, along with other miscellaneous applications of Deep Neural Networks.

6.5 Artificial Intelligence Challenges

The big challenge of Artificial Intelligence is that it learns from the data, and any inaccuracies in the data are reflected in the results. AI systems are very specialized and are built just to perform a single task and keep learning to become better and better at it. The process these systems follow is to look at the inputs given and the results produced. It looks at the best result produced and notes down those input values. The ethical challenges of Artificial Intelligence are shown in Figure 6.3.

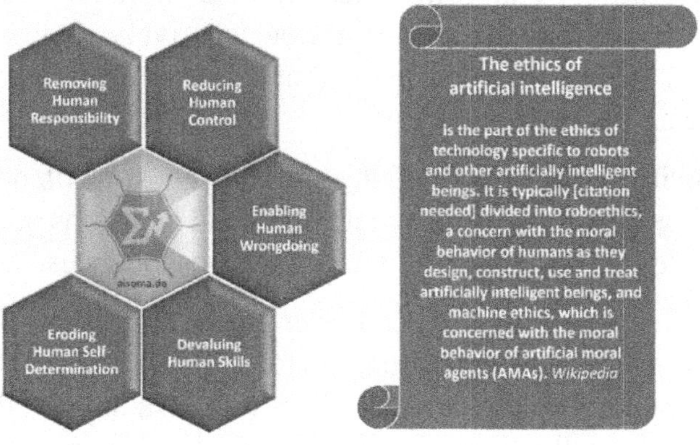

Figure 6.3: Ethical Challenges of Artificial Intelligence

Organizations have access to more data than ever before at the present time. However, datasets that are applicable to Artificial Intelligence applications to learn are rare. The most powerful AI machines are those that are trained in supervised learning. This kind of training requires labeled data. Labeled data is organized to make it understandable for machines to learn. Organizations are investing in design methodologies and focusing on how to create Artificial Intelligence models despite the scarcity of labeled data.

Most of the AI applications are based on massive volumes of data to learn and make intelligent decisions. Machine learning systems depend on data that is often sensitive and personal in nature. These systems learn from the data and improve themselves. Due to this systematic learning, these systems can become prone to data breaches and identity theft. Also, the quality of some of the decisions made by AI systems depends on data, which is often associated with ethnic, communal, gender, or racial biases. Based on a bias hidden in the algorithms, proprietary algorithms make crucial decisions that could lead to unethical and unfair results. Hence, the urgent need for organizations working on AI is to train AI systems with unbiased data and create algorithms that can be easily explained. AI challenges can be summarized as follows:

- **AI Integration:** The compatibility of current programs with AI requirements and implementation of AI into these programs without stopping current output should be ensured. The AI interface needs to be set up so that infrastructure, data storage, and data input are considered, and the output is not negatively affected. Additionally, once this is completed, ensure that all personnel are trained on the new system.
- **Algorithm Bias:** Bias is one of the biggest challenges facing AI. If the bias in the algorithms that make vital decisions goes unrecognized, it could lead to unethical and unfair consequences.
- **Data Integration:** Data can be collected on various applications with a multitude of formats, such as text, audio, images, and videos. The wide range of platforms to collect this data adds to the challenges of artificial intelligence. To succeed, all this data must be integrated so that the AI can understand and transform it into useful results.

- **Strategic Implementation:** To be successful, a strategic approach needs to be established while implementing AI. This includes identifying areas that need improvement, setting objectives with clearly defined benefits, and ensuring a continuous process improvement feedback loop.
- **Legal Concerns:** One of the challenges of AI includes legal concerns about collecting sensitive data that might violate laws. Even if it is not illegal, organizations need to be careful of any perceived impact that might have a negative impact. If the public perceives the data collected as violating their data privacy, it might not be worth the potential public relations backlash.

6.6 Impact of Artificial Intelligence on Jobs

The possibility of Artificially Intelligent systems replacing much of modern manual labor is extraordinarily strong soon. While AI will not replace all jobs, what is certain is that AI will change the nature of work, with the only question being how rapidly and how profoundly automation will alter the workplace.

The evidence of which jobs will be supplanted is starting to emerge. Amazon has just launched Amazon Go, a cashier-free supermarket in Seattle where customers just take items from the shelves and walk out. What this means for the more than three million people in the US who work as cashiers remains to be seen. Amazon is leading the way in using robots to improve efficiency inside its warehouses. These robots carry shelves of products to human pickers who select items to be sent out. Amazon has more than 100,000 bots in its fulfillment centers, with plans to add many more. But Amazon also stresses that as the number of bots has grown, so has the number of human workers in these warehouses. However, Amazon and small robotics firms are working on automating the remaining manual jobs in the warehouse, so it's not a given that manual and robotic labor will continue to grow.

Fully autonomous self-driving vehicles aren't a reality yet, but by some predictions, the self-driving trucking industry alone is poised to take over 1.7 million jobs in the next decade, even without considering the impact on couriers and taxi drivers. Millions of people are working in administration, entering and copying data between systems, and chasing and booking appointments for companies. As software gets better at automatically updating systems and flagging the information that is important, the need for administrators will fall. As with every technological shift, new jobs will be created to replace those lost. However, what's uncertain is whether these new roles will be created rapidly enough to offer employment to those displaced and whether the newly unemployed will have the necessary skills or temperament to fill these emerging roles.

Because AI systems must be continually recalibrated and trained, there will always be a 'venture' element in any AI project because the human mind (and emulates it) can be unpredictable. This uncertainty must be planned for in any AI ROI formula. One step AI promoters can take is to educate upper management on the risks so that these risks can be planned for and managed. A second step is to factor risk into the ROI formula by adding a 20 percent cushion to AI project cost projections as a margin for the unknown and the unexpected. The impact of AI on employment has been much heralded across all industries. The Financial Services sector is expected to be one of those most impacted soon. The employment impact of automation and AI on Financial Services is expected to be the greatest of all industries into the late 2020s, with only the transport industry experiencing a greater impact in the long term.

Introducing AI in a company is going to impact systems and business processes. Minimally, systems that need to communicate and exchange information with AI must be integrated with the AI. System processes will, by necessity, be modified. As AI is rolled out to business operations, it will undertake processes that were formerly performed by humans, causing the displaced humans to enter new and/or revised job roles.

6.7 Conclusion

Artificial Intelligence is far better at predicting behavior than the human mind. AI can also develop value systems, leading to questionable decisions. Artificial Intelligence and its subcomponents (machine learning, cognitive computing, and even deep learning) have grown in interest as data size and diversity continue to grow and the cloud becomes a popular option for quickly and economically scaling compute power and data storage. Artificial Intelligence can be used to enhance applied analytics to help organizations better understand their customers and deliver a great customer experience. It also provides practical advice on how organizations can use what they may already be doing to become even more effective in marketing and other business processes.

Algorithms are, of course, the key technical feature of most AI systems, and it is impossible to separate data from algorithms since machine learning algorithms learn from data. In fact, the greatest impediment to effective algorithms is insufficient, poor quality, or unlabeled data. The absence of a data-driven culture affects AI as much as any 0. It means that the company and its leaders are unlikely to be motivated or knowledgeable about AI and, hence, unlikely to build the necessary AI capabilities to succeed. Even if AI applications are successfully developed, these may not be broadly implemented or adopted by users. In addition to culture, AI systems may be a poor fit for organizational structure, strategy, or poorly executed change management. In short, the organizational and cultural dimension is critical for any firm seeking to achieve a return on AI.

Organizational leaders are contemplating the evolution of data-driven decision-making to the automated decisions that define artificial intelligence. The major considerations range from data preparation through experimentation and model development to implementation and monitoring. An analytically driven operating environment leads to the best automated models that realize the value-creation possibilities of artificial intelligence.

CHAPTER 7: APPLICATIONS OF ARTIFICIAL INTELLIGENCE IN SHARED SERVICES

Artificial Intelligence is growing in its applications, some even with impressive capabilities such as self-driving vehicles or social humanoids. Though we do not currently have a fully functional AI-based robot, it seems just a matter of time before we will have a fully functional humanoid capable of working with us or supporting us in our homes. Here are a few applications that we regularly use that are extremely important in developing any technology that can engage in the same way humans can in real life. Natural Language Generation is a sub-discipline of Artificial Intelligence that recognizes human language in text form and converts it into a form that enables the computer system to communicate the data efficiently and accurately. For example, it is used to take raw market data and generate reports of market summaries. It can also be used to generate short pieces of text in interactive conversations (a chatbot), which a text-to-speech system might even read out.

7.1 Natural Language Processing (NLP)

Natural Language Processing is an application of Artificial Intelligence that enables machines to both process and comprehend human language in the way it is written. In recent times, deep learning models can not only analyze large volumes of text but can also come up with services such as those providing a summarization of text, language translation, context modeling, and even sentiment analysis. Natural language understanding is a branch of NLP that converts the natural language spoken by humans into structured data. It can perform two tasks - intent classification and entity extraction. A user's intent is defined as what a user is trying to convey or accomplish. For example, when we read a sentence, we immediately understand the meaning or intent behind that sentence. Intent classification is a two-step process. First, we feed the model with labeled data that provides the list of known intents and example sentences that correspond to those intents. Once trained, the model can classify a new sentence that it feels fits wells into one of the predefined intents.

An important use of Natural Language Generation (NLG) is the classification of text. This is specifically used to control spam. Several email providers use NLG-based services to analyze the content of the emails to understand whether the mail is genuine or fake. NLG can help to do the following:

- Manually search data, comparing or scanning tables
- Analyzing charts, metrics, or maps
- Developing recurring status reports.

7.2 Conversational AI

Conversational AI is a set of technologies that enable computers to simulate conversations and refers to the use of messaging apps, speech-based assistants, and chatbots to automate communication and create personalized customer experiences at scale. This form of AI is preferred because it is time-bound and engaging. In business, the element of conversation keeps the customer engaged whilst, in the background, it can simultaneously work on the customer satisfaction aspect of the transaction. The number of queries that a business can handle can be quite large, and as such, the use of Conversational AI can reduce the query resolution time, and queries can be solved effectively.

The applications for artificial intelligence are endless. Technology can be applied to many different sectors and industries. AI is being tested and used in the healthcare industry for suggesting drug dosages, identifying treatments, and aiding in surgical procedures in the operating room.

7.3 Principles of AI

Seven principles should always be kept in mind while adopting AI (Upadhyay, 2022):

1. Have all data in one place or have them seamlessly connected to one system.
2. As a first step, break down the core problem into specific use cases that AI may or may not solve.
3. Choose the software that's the right fit for your needs, budget & existing organizational systems and processes, rather than going for the most popular ones.
4. Choose AI software that can show the rationale behind its analysis, especially for critical tasks and decision-making.
5. Ensure that data is proper and ready for AI use.
6. Effective AI requires proper adoption by the users, the right processes to support it, the right measures to keep it working properly, and only the desired degree of disruption to existing systems and processes.
7. Not all solutions have to be AI.

These principles are designed to aid the preparation for AI use and ensure its effectiveness. Principle #5 refers to proper data, while Principles #2, #3, and #7 correspond to the choice of AI solution to solve a business problem.

7.4 Risks with AI

Without a proper policy in place, organizations are subject to three broad areas of risk. While not a comprehensive list, these include:

- Systems and Socioeconomic Risk
- Privacy and Security Risk
- Legal and Financial Risk

The point to note here is this: Each of these risks comes with associated dilemmas facing us today. The right approach to managing them often boils down to what the right balance of ethics and practicality is. That balance itself tends to shift when we have personal stakes in play (Upadhyay, 2022).

7.4.1 Systems and Socioeconomic Risk

Systems and socioeconomic risk primarily pertain to the risk of replacing existing systems and resources. Any new technological innovation that transforms a socioeconomic cycle results in the extinction of processes and jobs that serve older technologies and systems, which are replaced by newer processes and jobs based on the requirements to support the new system. A complete overhaul of an organization is rarely advisable, regardless of how often we may feel tempted to do so. Managers must protect themselves from the fallacy of committing to a new shiny object by disregarding what they already have. In fact, in almost every consulting project around AI enablement and use, there is a need to start with the basics to get the data dictionary in place with traditional processes, data sources, owners, and workflows having to be properly mapped out. Any new AI solution can then be designed to utilize what is in place while causing minimal intrusion as far as feasible.

Like any new technological insurgency in an economy over the centuries, the risk posed by AI on the socioeconomic structure is about a balance between the jobs it eliminates and the ones it creates and about the general levels of well-being it offers. Any problem had to be broken down into individual tasks or use cases that AI may solve. Likewise, it is the tasks within a job that AI replaces. The overall impact on any job, therefore, depends on the number of tasks handled by that job that AI can reliably replace. Any rule-based task is potentially replaceable by AI as it matures. These tasks are performed per set rules and procedures in a largely predictable and unchanging set of scenarios.

On the other hand, more interpersonal tasks, operating in unpredictable scenarios or involving critical decision-making that may not abide by strict rules, will require humans for the foreseeable future. The dilemma is that employing AI allows organizations to cut costs and sustain themselves financially so that they and all other jobs can survive. It also allows organizations to remove any inherent discrimination and inefficiencies that humans may bring. Moreover, the overuse of AI can blind a company to biases in the AI model's decision-making and, of course, rob it of the power to make intuitive or critical contextual decisions, costing it dearly.

7.4.2 Privacy and Security Risk

AI can easily and illegally breach customers' and users' private information. It can also use approaches or make either unethical or biased predictions. A system's risk can be seen as a security risk because an AI system can easily act unfairly against those it is biased against, whether it is one individual or a whole community. A person has the right to own and control the use of data related to him or her. The intention is not necessarily evil, as sometimes, the very nature of AI development is to solve a problem on the edge of risk. Fake news is a big problem, after all. What we must guard against is the misuse of such technology not only by external sources but also by the employees within, which can compromise an organization's security.

7.4.3 Legal and Financial Risk

Legal and financial risk has largely to do with repercussions of AI misuse - intentionally or unintentionally, particularly with black-box AI. With sound policies in place, it is important that we opt for transparent AI solutions and not black boxes. The more automated and self-learning AI becomes, the less visibility we have into how it is analyzing something. AI can afford to be a black box if it is dealing with a relatively simple task in that context. After all, while AI can identify the presence of skills or look for success indicators in the wording on a resume, it cannot beat the depth of assessment we can make by physically interacting with the candidates and looking at their body language. On the other hand, if AI were not involved at all, recruiters would have a tough time going through hundreds of resumes for a job. They would likely make an even narrower or even more biased assessment than the AI would have in terms of that initial filtering. It is a perfect example of why the best and most secure business solutions today are derived from people and AI working together.

AI's job is to enhance the productivity of humans. It can show us things we couldn't see otherwise and make a lot of our tasks easier and quicker. But it cannot be the sole decision-maker. The final decision should always rest with humans if an organization wishes to avoid falling into legal trouble or suffering unforeseen financial losses besides going ethically astray. Human intervention will always remain important.

7.5 Applications of AI

Artificial intelligence is widely used in business applications, including automation, data analytics, and natural language processing. Across industries, these three fields of AI are streamlining operations and improving efficiencies. Automation alleviates repetitive or even dangerous tasks. Data analytics provides businesses with insights never before possible. Natural language processing allows for intelligent search engines, helpful chatbots, and better accessibility for people who are visually impaired. Other common uses for AI in business include:

- Transferring and cross-referencing data; updating files.
- Consumer behavior forecasting and product recommendations.
- Fraud detection
- Personalized advertising and marketing messaging
- Customer service via telephone or chatbots

Artificial Intelligence has marked its presence in almost every industry and walk of life. It has not only reduced human interventions in various operations but also helped humans do their jobs better. Fields like social media, Consumer Electronics, Robotics, Travel and Transportation, Finance, Healthcare, Security, Surveillance, E-commerce, etc. are already benefiting from AI. Digital Marketing and AI go together. In digital marketing, there is a massive requirement to process tons of data. Artificial Intelligence helps digital marketers process data faster, which allows them to create digital strategies more efficiently. The capabilities of AI in Digital Marketing are massive. Figure 7.1 shows applications of Artificial Intelligence in Digital Marketing.

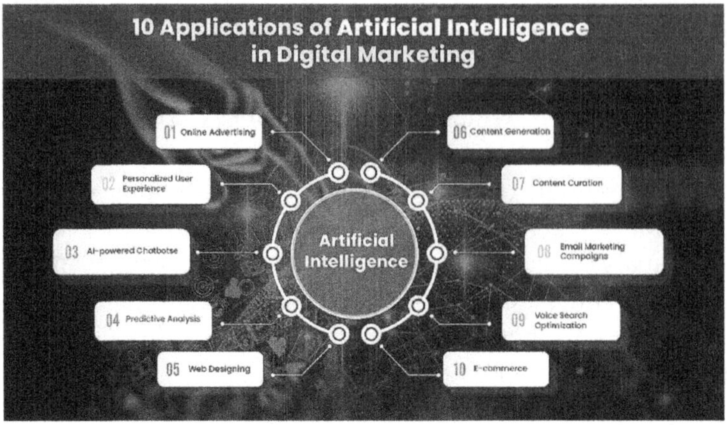

Figure 7.1: Applications of Artificial Intelligence in Digital Marketing

Maintaining equipment requires regular repairs and upkeep. AI provides a way to know exactly when to conduct these tasks and save money while cutting downtime. AI collects information from equipment sensors, combining it with maintenance records. The system analyzes the information to predict the best time to repair the equipment, which is called predictive maintenance.

One of the hardest aspects of improving inventory levels is predicting what to keep in stock. AI is used to predict how much stock is needed to decrease inventory. Cutting needless inventory is the best way to free up warehouse space to stock more products being sold. Predictive modeling uses AI to examine consumer trends and help identify which products to stock. Revenue losses can be reduced by not having the right products immediately available.

There are too many variables at play for humans to predict the appropriate inventory levels in real-time accurately. AI can solve these inaccuracies, making the facilities more efficient. Reviewing contracts and comparing them against previous versions and standard templates is a significant workload. Tight timelines and volume of work can lead to poor contract language and both lost value and increased risk. AI can be applied to extract data from complex unstructured documents to classify elements and to understand automated business processes and tables.

Artificial Intelligence allows for synchronization throughout a supply chain, which helps in sending information about predicted customer demand down the line so that the suppliers produce just enough to avoid overproduction waste. The connectivity of devices and sharing data is critical for AI use. While this connectivity allows equipment to send information about production back and forth, it can also be useful throughout the supply chain. The collected information can be stored through AI in the cloud for others to access. Machines, having AI capabilities, can mimic humans, automate manual tasks, and learn on the go, just like humans. With automation coming into the picture, repetitive and time-consuming tasks go into the plate of AI. AI-powered systems exhibit human intelligence and learn with time, indicating that these machines can eventually conduct critical-thinking jobs and make decisions themselves. Realizing this unique potential of AI, businesses in the transportation sector are making significant investments to improve revenue generation and stay ahead of their competitors.

7.6 Applications of AI Systems

There are numerous real-world applications of AI systems. Below are some of the most common use cases (IBM, What is Artificial Intelligence (AI)?, 2023):

- **Speech recognition:** It is also known as automatic speech recognition (ASR), computer speech recognition, or speech-to-text, and it is a capability that uses natural language processing (NLP) to process human speech into a written format. Many mobile devices incorporate speech recognition into their systems to conduct voice searches—e.g., Siri—or provide more accessibility around texting.
- **Customer service:** Online virtual agents are replacing human agents along the customer journey. They answer frequently asked questions (FAQs) around topics like shipping, providing personalized advice, cross-selling products, suggesting sizes for users, and changing the way we think about customer engagement across websites and social media platforms. Examples include messaging bots on e-commerce sites with virtual agents, messaging apps like Slack and Facebook Messenger, and tasks usually done by virtual assistants and voice assistants.
- **Computer vision:** This AI technology enables computers and systems to derive meaningful information from digital images, videos, and other visual inputs, and based on those inputs, it can act. This

ability to provide recommendations distinguishes it from image recognition tasks. Powered by convolutional neural networks, computer vision has applications within photo tagging in social media, radiology imaging in healthcare, and self-driving cars within the automotive industry.

- **Recommendation engines:** Using past consumption behavior data, AI algorithms can help discover data trends that can be used to develop more effective cross-selling strategies. This is used to make relevant add-on recommendations to customers during the checkout process for online retailers.
- **Automated stock trading:** Designed to optimize stock portfolios, AI-driven high-frequency trading platforms make thousands or even millions of trades per day without human intervention.

7.7 Uses of AI

The enterprise's first and most obvious uses for AI often revolve around the bottom line, i.e., cost savings and efficiency. For example, applications of AI can be used to improve email spam filters and web protections. The AI examples in industrial settings are warehouse robots that automatically pull products from shelves for shipping. This use of AI helps protect human workers from the dangerous tasks around a warehouse. It also improves accuracy and safety in the warehouse. Smart robots can avoid obstacles like people and shelving for easier maneuvering when retrieving or storing products. With AI, multiple pieces of equipment can be controlled in facilities from a single panel, which lessens the need for numerous low-skilled workers. Figure 7.2 shows the uses of Artificial Intelligence in business.

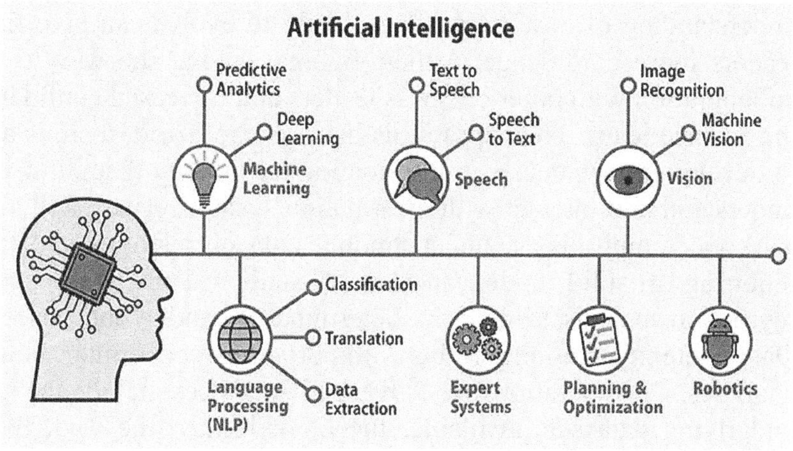

Figure 7.2: Uses of Artificial Intelligence in Business

Artificial Intelligence is changing the way business is done in several industries, from improving access to medical care in the healthcare field to detecting fraud in financial institutions, improving traffic management systems in transportation, helping companies run more efficiently in technology, and mitigating customer service issues in retail.

7.8 Advanced Analytics

a. The Finance functions leverage analytics to provide financial insight to business units by helping them understand what impacted their financial results. Financial planning and analysis use historical data and assumptions to model and forecast financial results. With advancements in machine learning algorithms, AI has enabled advanced business analytics. Predictive analytics uses data, algorithms, and machine-learning techniques to anticipate future outcomes. Prescriptive analytics takes this one step further by guiding predictions into actions. The role of finance must evolve to do more than report historical key performance indicators and make periodic forecasts. Businesses are moving faster than ever before, and the speed of insight for finance must accelerate as well. The leveraging of Big Data by advanced analytics solutions can help develop a deeper understanding of new market trends, identify new indicators for performance management, and improve the accuracy and timeliness of the forecasting process. More effective forecasting and a deeper understanding of how markets are likely to evolve can give finance leaders more confidence in their figures, paving the way to better collaboration with other business leaders and increased confidence in the finance team. To adapt to this data-centric world, there is a need to develop competencies in information technology that allow one to understand and interact with information systems. There will also be a heavier emphasis on understanding data correlations and trends, requiring a solid understanding of statistical methods such as regression analysis, sample size determination, and hypothesis testing. Data contextualization will be as important as performing the actual analytics. Considerations for Advanced Analytics include the size of underlying data sets available; the more limited the data, the less reliable the prescribed outcome; quality of data; a direct correlation between data quality and reliability of predictive outcomes, the existence of potential bias in underlying data and assumptions that may skew outcomes (e.g., selection bias, sampling bias, modeling bias, etc.) and availability of qualified data scientists.

The benefits of Advanced Analytics include:

- Fact and data-driven recommendations and outcomes instead of gut instinct
- Support for analyzing vast amounts of internal and external data, including real-time events, to identify otherwise hidden correlations and trends
- Reduction of time to insight and enabling of faster decision-making
- Develop a deeper understanding of business and market trends.

7.9 Applications of Artificial Intelligence in Shared Services

Artificial intelligence has many applications in Shared Services organizations where it is used to automate processes. Figure 7.3 shows candidates for application of Artificial Intelligence in Shared Services.

Finance Shared Services	HR Shared Services	Procurement Shared Services	IT Shared Services
Invoice Processing	Payroll	Purchasing	Ticket Management
Accounting	Selection & hiring		Database Management
General Ledger			Call Centre
Financial Analysis			
Reporting			

Figure 7.3: Candidates for Application of Artificial Intelligence in Shared Services

As an example, one of the biggest challenges in processing invoices is the lack of uniformity. Invoicing and payment processing require multiple fields, and companies and employees are often inconsistent in how invoices are formatted. That's called "unstructured data," which is time-consuming and error-inducing. There is a need to translate unstructured data into organized, actionable, and reliable data that is easily accessible to drive informed business decisions. AI learns from every human transaction and changes and adapts as follows:

- Provides dramatic cost savings.
- It is guaranteed to meet or exceed your ROI goals.

- Allows for automated end-to-end processes, which offer higher straight-through processing.
- Is simple to use, even for non-data scientists.
- Eliminates human error from manually processed invoices.
- Allows for fast, typically one-call, issue resolution, which leads to better vendor relations.
- Easily integrates with RPA and other process automation tools.
- Is flexible, scalable, and easily customizable.
- Processes any data type, including text, documents, images, audio, and video.

7.10 Conclusion

Within the Finance function, AI can be applied to replace repetitive and labor-intensive tasks, performing such transactional work with increased speed and accuracy. Moreover, with its capacity to learn from large data sets, the technology can also improve accuracy in budgeting and forecasting to enhance companywide decision-making. Here are three examples (Troitsky & Tay, 2022):

1. **Financial Planning**: AI can help identify critical performance drivers, improve forecasting outcomes, and increase speeds to execute an accurate forecast. AI can also eliminate the human bias that goes into a forecast. This doesn't mean that humans are cut out of the process; rather, it means that a solid baseline can be produced for Finance staff to review, tweak, and adjust drivers if needed. AI leverages financial, operational, and third-party data and learns from patterns to apply reason to future projections.
2. **Financial Close:** AI can receive thousands of transactions and make decisions on how to categorize them: whether they should be booked immediately or spread over multiple periods. AI can also automatically make required journal entries, generate financial statements, and provide commentary on how the company performed relative to budget and relative to the prior year. AI can eliminate bottlenecks and do the grunt work, so Finance teams just have to

validate it. AI can help eliminate the scramble to close the books on time and without errors, making it a less tedious and stressful process.

3. **Financial Risk Sensing**: AI models can analyze customer data, look for behavior patterns in transactions, and flag those that fall outside certain parameters. This learning from historical data can be applied for the future, providing foresight to potential risks. AI can be a powerful tool for the risk and compliance requirements of the Finance function.

- AI can help Finance departments shift human focus to activities that add real strategic value. The foundation of a good result for AI in Finance is to build cross-functional teams for AI. Many companies are creating a Center of Excellence for AI applications. The benefits of using AI in Finance function can be numerous:
- Cost reduction
- Speed to execution
- Reduced complexity
- Transformed engagement.
- Fueled innovation
- Fortified trust

The Finance function can be a transformative agent of innovation by using AI. The results cannot only provide the Finance team with better, faster information, but they can also influence the entire company's strategic thinking.

CHAPTER 8: GENERATIVE ARTIFICIAL INTELLIGENCE

Generative Artificial Intelligence (AI) is a type of artificial intelligence that enables machines to generate added content or data, such as text, images, or even entire applications. Unlike other types of AI, such as predictive or prescriptive AI, which are based on existing data, generative AI creates new data or content by learning patterns and rules from existing data. Generative AI solutions such as ChatGPT have exploded on the scene in recent months. AI will not replace the Finance function, but the Finance leaders who use generative AI solutions effectively will have a competitive edge (Beebe, 2023). Generative AI is based on machine learning algorithms that use deep neural networks to analyze and learn from substantial amounts of data. These algorithms can generate added content by synthesizing patterns and relationships in the data and creating new outputs that match these patterns. This process is known as "generative modeling."

8.1 What is Generative AI?

Generative AI, or "Generative Artificial intelligence" to give its full name, is a type of artificial intelligence technology that uses machine learning algorithms to create new and original content like images, videos, text, and audio.

- It uses neural networks to identify patterns and structures within the existing data to generate new and entirely unique content.
- Does this without specific instructions.
- Simulates complex events, predicts outcomes, and produces high-quality content that rivals human creativity.
- A machine learning process called deep learning uses interconnected nodes or neurons in a layered structure that resembles how the human brain works.
- Traditional AI excels at pattern recognition; Generative AI excels at pattern creation.

A Generative AI Centre of Excellence (CoE) serves as the nucleus of an organization's AI capabilities, fostering an environment of innovation and continuous learning. It ensures that the adoption and integration of generative AI aligns with the organization's strategic objectives, ensuring maximum value and impact.

8.2 How does Generative AI work?

Generative AI is a type of artificial intelligence that generates new content by modeling data features from large datasets fed into the model. While traditional AI systems can recognize patterns or classify existing content, generative AI can create new content in many forms, including text, image, audio, or software code. Generative AI is both a transformative and disruptive technology that may significantly alter the way consumers, industries, or businesses operate. It has the potential to enable creativity and innovation that could improve services and business operations. Some of the useful applications of generative AI are in the following areas:

- **Healthcare:** Assists healthcare providers to make faster diagnoses. AI holds the promise to make personalized treatment plans commonplace. It can be leveraged to enable therapeutic targets and novel drug candidates.
- **Software development:** Enables software developers to generate code, assist in debugging, or offer code snippets. This can potentially help speed up the development and release of software products.
- **Online marketplace:** Generates human-like responses in a chatbot and conversational agents, which can help organizations improve customer service and reduce support costs.
- **Business:** Creates personalized customer communications for existing and prospective clients, as well as generates predictive sales modeling to forecast customer behavior.
- **Publishing and media:** Enables content creators to produce unique outputs for use in marketing campaigns, advertising, television, and video productions. On-demand content can be generated quickly and with fewer resources, leading to significant cost reduction.

- **Education:** Allows educators to create personalized learning plans tailored to students' individual performance, needs, and interests. This could help teachers better support their students.
- **Cyber security:** Facilitates enhancement of cyber defense tools against ransomware and other attacks. Assists cyber security practitioners in scanning large datasets more easily to identify potential threats and minimize false positives by filtering out non-malicious activities.

Generative AI works by training machine learning models on large datasets of existing content or data. The machine learning models learn patterns and relationships in the data and use these patterns to generate added content that is like the original data. For example, a generative AI model trained on a dataset of images of cats can generate new images of cats that have features and characteristics similar to the original dataset. Similarly, a generative AI model trained on a dataset of customer service conversations can generate new conversations that sound like human agents spoke them. Generative AI can be trained on a variety of data types, including text, images, video, and audio. The type of data that is used to train the generative AI model will determine the type of content or data that the model can generate.

8.3 Applications of Generative AI in Shared Services

Generative AI has a wide range of potential applications in Shared Services. Here are some of the ways that generative AI can be used to improve efficiency and productivity in shared services:

- **Customer Service:** Generative AI can be used to create chatbots that can manage routine customer service inquiries, freeing up human agents to focus on more complex issues. Chatbots powered by generative AI can generate responses like human agents, providing a seamless customer experience.
- **Finance and Accounting:** Generative AI can be used to automate repetitive and time-consuming tasks such as data entry, account reconciliation, and invoice processing. This can free up employees in shared services to focus on higher-value activities such as financial analysis and strategy.
- **Marketing and Sales:** Generative AI can create personalized content and customer experiences, such as customized product recommendations or marketing messages. This can improve customer satisfaction and loyalty, which can benefit shared services such as marketing and sales.
- **Human Resources:** Generative AI can be used to streamline recruitment processes by generating job descriptions, screening resumes, and scheduling interviews. This can save time and resources for HR teams, allowing them to focus on more strategic initiatives such as employee engagement and retention.
- **Legal:** Generative AI can be used to assist with legal research and document drafting. For example, generative AI can be trained on a dataset of legal briefs and use this data to generate new briefs that are relevant to a specific case. This can save time and resources for legal teams, enabling them to focus on more complex legal issues.

Figure 8.1 summarizes the use of Generative AI in Shared Services Processes (Chazey, Unleashing the Power of Generative AI in GBS, 2023).

Human Resources	• Policy Drafting: ChatGPT can be used to draft documentation, employee handbooks, or policies. • Learning Customization: Generative AI can customize learning materials and adapt them to your audience. • Summarize Feedback: It can be used to summarize employee feedback. • Interview Questions: Draft questions to interview specific skills. • Employee Recognition: Draft personalized appreciation messages for employee achievements.
Procure to Pay	• Automated invoice processing: AI models can extract relevant information from invoices, verify accuracy, and automate the approval and payment processes. • Spend analysis: AI algorithms can analyze spending patterns, identify cost-saving opportunities, and provide recommendations for optimizing procurement strategies. • Supplier risk assessment: Generative AI can analyze supplier data, financial reports, and market trends to assess supplier risks and predict potential disruptions.
Record to Report	• Financial report generation: AI models can generate financial reports, including balance sheets, income statements, and cash flow statements, based on input data and accounting rules. • Fraud detection: AI algorithms can analyze financial transactions, identify suspicious patterns, and flag potential fraud cases for investigation. • Account reconciliation: Generative AI can automate reconciliation by matching transactions, identifying discrepancies, and suggesting corrective actions.
Order to Cash	• Sales forecasting: AI models can analyze historical sales data, market trends, and external factors to generate accurate sales forecasts, aiding demand planning and resource allocation. • Credit risk assessment: AI algorithms can assess customer credit by analyzing historical payments, financial data, and industry information. • Order management optimization: Generative AI can optimize order fulfillment processes by predicting demand, optimizing inventory levels, and dynamically adjusting production schedules.

Figure 8.1: Use of Generative AI in Shared Services Processes

8.4 Benefits of Generative AI in Shared Services

There are several benefits of using generative AI in shared services, including:

- **Increased efficiency and productivity:** Generative AI can automate repetitive and time-consuming tasks, freeing employees in shared services to focus on higher-value activities. For example, chatbots powered by generative AI can manage routine customer service inquiries, leaving human agents to address more complex issues.
- **Improved accuracy and quality:** Generative AI can analyze substantial amounts of data to identify patterns and insights that humans may miss. This can lead to more accurate and higher-quality results in shared services, such as finance and accounting.
- **Enhanced customer experience:** Generative AI can create personalized content and experiences for customers, such as customized product recommendations or marketing messages. This can improve customer satisfaction and loyalty, which can benefit shared services such as marketing and sales.
- **Faster innovation:** Generative AI can generate ideas and prototypes faster than humans, enabling shared services to innovate and develop new products or services more quickly. It can give organizations a competitive advantage in their respective markets.
- **Cost savings:** Generative AI can reduce costs in shared services by automating tasks and improving efficiency. This can free up resources that can be allocated to other areas of the organization.

8.5 Drawbacks of Generative AI in Shared Services

Whilst there are several benefits, there are also potential roadblocks to using this new concept in shared services, including:

- **Job Displacement:** Generative AI has the potential to automate many tasks currently performed by humans in shared services. This could lead to job displacement and require re-skilling or up-skilling of employees.
- **Security Risks:** Generative AI can be vulnerable to security risks such as cyberattacks or data breaches. This can pose a risk to sensitive data managed by shared services such as HR or legal.
- **Bias:** Generative AI is only as unbiased as the data it is trained on. The resulting output from generative AI can also contain biases if the data contains biases. This can create ethical and legal issues for shared services such as hiring or promotions.
- **Lack of Human Judgment:** Generative AI uses algorithms and data to generate content or insights. This can limit the ability of shared services to exercise human judgment in decision-making processes.
- **Reliance on Technology Vendors:** Generative AI requires specialized expertise and infrastructure to implement and maintain. Shared services may become reliant on technology vendors to provide these services, creating potential vendor lock-in and reducing flexibility.

Generative AI has the potential to significantly impact shared services and change our landscape forever. While the benefits of improved efficiency, accuracy, and innovation are compelling, shared services should be aware of the risks, such as job displacement, security, and bias. Shared services can harness their potential while mitigating their risks by carefully managing the implementation and use of generative AI.

8.6 Risks of Generative AI

While the capabilities of generative AI technology present great opportunities, they also bring many concerns. Generative AI can enable threat actors to develop malicious exploits and potentially conduct more effective cyber-attacks. A huge concern is that it can provide threat actors with great powers to influence. For example, deliberate manipulation of the underlying code and the tools using it can introduce supply chain risks from insider threats at the design level to the distribution and patching of software (Cybersecurity, 2023). Here are some of the potential risks:

- **Misinformation and Disinformation:** Content may not clearly be identified as being AI-generated and could potentially lead to confusion (misinformation) or deception (disinformation). Threat actors can use this in scams or fraudulent campaigns against individuals and organizations.
- **Phishing:** Threat actors can craft targeted spear phishing attacks more frequently, automatically, and with a higher level of sophistication. Highly realistic phishing emails or scam messages could lead to identity theft, financial fraud, or other forms of cybercrime.
- **Privacy of Data:** Users may unknowingly provide sensitive corporate data or personally identifiable information (PII) in their queries and prompts. Threat actors could harvest this sensitive information to impersonate individuals or spread false information.
- **Malicious Code:** Technically skilled threat actors can overcome restrictions within the generative AI tools to create malware for use in a targeted cyber-attack. Those with little or no coding experience can use generative AI to easily write functional malware that could cause a nuisance to a business or organization.
- **Buggy Code:** Software developers may deliberately or inadvertently introduce unsecured and buggy code to the development pipeline. This could happen if they omit or improperly implement error handling and security checks.
- **Poisoned Datasets:** Threat actors can inject malicious code into the dataset used to train the generative AI system. This could undermine the accuracy and quality of the generated data. It could also increase the potential for large-scale supply-chain attacks.

- **Biased Content:** A majority of the training dataset fed into the LLMs comes from the open Internet. As such, generated content has a fundamental bias in that only limited amounts of the world's total data are online and available for AI to use. Also, generated content may be prejudiced if the training dataset lacks a balanced representation of data points.
- **Loss of Intellectual Property (IP):** Generative AI tools may enable sophisticated threat actors to steal corporate data faster more easily, and in bulk. Loss of IP (e.g., proprietary business information, copyrighted data, software code, or drug trial data) can devastate your organization's reputation, revenue, and future growth.

8.7 Conclusion

The primary way in which generative AI differs from other forms of AI or analytics is that it can generate new content efficiently, often in "unstructured" forms (for example, written text or images) that aren't naturally represented in tables with rows and columns. The underlying model that enables generative AI to work is called a foundation model. Some characteristics that set foundation models apart from deep learning models include that these models can be trained on extremely large and varied sets of unstructured data.

Deep learning models often can perform only one task, such as either classifying objects in a photo or performing another function, such as making a prediction. In contrast, foundation models can perform both functions and generate content as well. Foundation models amass these capabilities by learning patterns and relationships from the broad training data they ingest, which, for example, enables them to predict the next word in a sentence.

Given the versatility of a foundation model, companies can use the same one to implement multiple business use cases; something rarely achieved using deep learning models. A foundation model incorporating information about a company's products could potentially be used both for answering customers' questions and supporting engineers in developing updated versions of the products. As a result, companies can stand up to applications and realize their benefits much faster.

CHAPTER 9: ChatGPT AND SHARED SERVICES

Shared Services organizations can use ChatGPT to create a series of personalized responses to internal customer queries, which can provide a positive and engaging experience. With its ability to learn from past interactions and its natural language processing capabilities, ChatGPT can adapt to internal customers' unique needs and preferences. ChatGPT can be considered for the customer service team. ChatGPT was not strictly built with customer service in mind. Still, its ability to generate human-like responses and creatively answer questions has made it of interest to Shared Services organizations.

When OpenAI released its flagship product, ChatGPT, in November 2022, it reached 1 million users in only four days and 100 million users in two months, achieving this milestone more than 75% faster than any other tech giant. ChatGPT has continued to evolve quickly, beginning as an AI chatbot that only responds to user prompts. Now, it includes features such as analyzing data and interpreting images, leaving everyone from high school students to C-suite executives scrambling to learn ChatGPT so they can use it in new and productive ways (Cromwell, Harvey, Hasse, & Gardner, 2023).

9.1 What is ChatGPT?

ChatGPT is a hyperintelligent computer program that can write novels, screenplays, PhD dissertations, and research papers, all in a way that is indistinguishable from humans. Microsoft founder Bill Gates says ChatGPT is as significant as the invention of the Internet and will "change our world." Google CEO Sundar Pichai has said, "AI is one of the most important things humanities is working on. It is more profound than [electricity or fire]." In March 2023, not long after ChatGPT stunned the business community, a new version, ChatGPT-4, debuted and was even more powerful than the previous version. The uses of ChatGPT and its implications are beyond enormous, which explains why ChatGPT was adopted by over one million people in just five days, the fastest rate of technological adoption in history.

ChatGPT stands for "Chat Generative Pre-Trained Transformer." The 'chat' naturally refers to the chatbot front-end that OpenAI has built for its GPT language model. The second and third words show that this model was created using 'generative pre-training,' which means it's been trained on huge amounts of text data to predict the next word in each sequence. The speed, scale, and scope of impact of ChatGPT are unique. But harnessing the power of emerging technology requires a new approach. Unlike traditional methods for innovation, such as design thinking, ChatGPT poses a fundamentally different challenge because there are no clear problems to address, user pain points to alleviate, or KPIs to achieve. In fact, ChatGPT can seemingly solve problems in almost any industry, domain, or context. Using it, therefore, requires a different skill: emergent thinking, which involves generating ideas for innovation without fully understanding the problem that needs to be solved.

ChatGPT is an AI chatbot that was initially built on a family of large language models (LLMs) collectively known as GPT-3. OpenAI has now announced that its next-gen GPT-4 models are available. These models can understand and generate human-like answers to text prompts because they have been trained on vast amounts of data. For example, ChatGPT's model was trained on text data from the internet, which OpenAI says included books, articles, websites, and even social media. Because it has been trained on hundreds of billions of words, ChatGPT can create responses that make it seem like, in its own words, "a friendly and intelligent robot." This ability to produce human-like and frequently accurate responses to a vast range of questions is why ChatGPT became the fastest-growing app of all time. The fact that it can also generate essays, articles, and poetry has only added to its appeal (and controversy in areas like education).

ChatGPT has been created with one main objective: to predict the next word in a sentence based on what typically happens in the gigabytes of text data that it's been trained on. Once a question or prompt is given to ChatGPT, it passes through the AI model, and the chatbot produces a response based on the information you've given and how that fits into its vast amount of training data. During this training, ChatGPT has learned what word, or sequence of words, typically follows the last one in each context.

But the short answer? ChatGPT works thanks to a combination of deep learning algorithms, a dash of natural language processing, and a generous dollop of generative pre-training, which all combine to help it produce disarmingly human-like responses to text questions. Even if it has only been trained to do is fill in the next word based on its experience of being the world's most voracious reader.

These activities tend to run counter to good business practices promoting efficiency and reliability, and they may even violate some of the core tenets of design thinking — namely, the need to identify a clear user problem to address before generating ideas for a solution. Yet, they're also critical when trying to leverage ChatGPT (or any other emerging technology, for that matter) for innovation.

Many companies have already started integrating ChatGPT into their products, processes, and services. By analyzing this burgeoning innovation landscape, we've identified three emergent-thinking pathways for leaders that can help increase the chances of implementing ChatGPT successfully while avoiding potential pitfalls.

9.2 Value Proposition of ChatGPT

Human Resources (HR) Shared Services play a critical role in managing a company's most important asset, its people. From recruiting new talent to managing employee benefits and compensation, HR Shared Services teams are responsible for ensuring the company's employees are engaged, productive, and motivated. HR Shared Services organizations can leverage ChatGPT to streamline their processes and achieve greater efficiency. ChatGPT can be a powerful tool for HR service providers in a variety of ways, including automating repetitive tasks, providing real-time support to employees, and enhancing the overall employee experience.

ChatGPT can automate repetitive tasks in the recruitment process, like screening resumes and scheduling interviews. Automating these types of tasks can free up HR service providers to focus on more strategic activities. ChatGPT can also assist in candidate engagement by providing real-time support and answering frequently asked questions about the company and the application process. HR service providers can also use ChatGPT to identify potential candidates who may be a good fit for the organization based on their skills, experience, and education.

HR service providers can use ChatGPT to improve the onboarding experience for new hires, reduce the workload on HR services team members, and ensure that new employees have access to the resources they need to be successful in their new roles. HR services teams can set up ChatGPT to provide real-time support and guidance to new hires, including answering common questions about company policies, procedures, and benefits and providing guidance on completing required paperwork. ChatGPT can also help automate administrative tasks, such as scheduling orientation sessions or sending reminders to new hires about required training. ChatGPT can provide employees with instant access to training materials and answer questions about workshops and programs. HR service providers can also automate administrative tasks with ChatGPT, including scheduling training sessions or providing reminders to employees about upcoming events. ChatGPT can even create personalized employee training plans based on their specific needs and skill sets.

To pursue the pathway for innovation, start by using ChatGPT without directing it toward any specific customer problems. This thought process can be challenging because most business training teaches us to be strategic thinkers and verify the importance of customer problems before spending any time developing solutions. However, starting with strategic goals like this can blind us to seeing ChatGPT's potential for new and valuable applications. Instead, start with an open mind and deeply engage with the tool to understand its core functions. Once you understand, natural human cognition will take over and start connecting to relevant problems.

For example, one of the core functions of ChatGPT is to interpret and generate natural language text that provides human-like responses to a wide variety of questions, prompts, and requests. Companies can use this function to enhance existing features of products or services that focus specifically on text-based interactions with customers. For instance, the product-management tool Notion now provides customers with immediate access to ChatGPT at the tap of the spacebar, which activates a prompt to receive an AI response that can be immediately integrated into existing tasks and workflows. As the functions of ChatGPT continue to grow, new possibilities will emerge for how it can be used to further enhance value propositions across a broader range of companies and contexts.

9.3 Expanding the Value Proposition of ChatGPT

ChatGPT expands a company's value proposition by solving new problems for customers in a way that complements its current offering. For emergent thinking, this first involves breaking down a value proposition into three core elements:

- The goals customers are trying to accomplish.
- The context in which customers are trying to accomplish those goals.
- The specific target demographic of the customers.

The next step is to assess how the core functions of ChatGPT can help expand each element in new directions. This definition and expansion process offers a complementary skill set to design thinking, which typically encourages people to think of many different solutions before deciding on one to implement. With emergent thinking, however, you apply those same brainstorming techniques to the problem rather than the solution.

During this process, it's key to maintain an open mind and avoid settling on a problem too quickly, which may be particularly challenging to do in conditions of uncertainty. ChatGPT has unleashed unprecedented pressure on companies to act quickly or suffer dramatic consequences for their future survival. However, research shows it's also possible to withstand these pressures and explore a wider range of innovation options. To do that, work with trusted collaborators so you can navigate ambiguity more effectively and take greater risks together.

9.4 New Value Proposition of ChatGPT

The final pathway uses ChatGPT to explore a new value proposition that customers may not even know they need yet. Although this may run counter to traditional methods for innovation, it also aligns with common advice for creating breakthroughs. A lot of times, people don't know what they want until we show it to them. "This wisdom seems to have been fully embraced by OpenAI, which released ChatGPT to the public and is now discovering countless ways it can be used in various contexts to accomplish new and unforeseen goals. This pathway for innovation represents the purest form of emergent thinking, which has the greatest uncertainty and must also be pursued with the greatest caution.

The main challenge comes when converging on a single problem to focus on out of the many dozens identified when brainstorming different goals, contexts, and target demographics. When companies have relatively clearer value propositions, such as increasing the speed of online ordering or improving the quality of online education, it can be easier to filter out problems that seem far-fetched or unlikely to reflect valuable customer needs. However, when customer needs are still unknown and emerging along with technology, it becomes difficult for teams to agree on a shared direction to pursue. This becomes especially challenging when trying to include the diverse perspectives that are fundamental to innovation because team members can pull the project in different directions despite their best intentions to collaborate.

To help overcome these challenges, identify a clear and coherent analogy that summarizes the core value proposition. Integrating many different features into a coherent product experience is important so developers and customers can easily understand its value. By creating an analogy for the product, we can avoid wasting valuable resources creating features that might not support the core value proposition, and you can help customers develop a clear mental model of the various features and why they can be useful.

ChatGPT is an unprecedented technology that has the potential to unlock value for millions of customers across many industries. However, pursuing these rewards also comes with considerable risks, as we're all entering uncharted territory with AI, and we must learn how to use it effectively to leverage it for positive rather than negative outcomes.

9.5 Shared Services Organizations and ChatGPT

With the right instructions and context, ChatGPT can be very useful to Shared Services organizations in doing the following:

- **Writing:** ChatGPT can be a good tool for writing articles. It will give mixed results if prompted to write a full article in one pass. But if we work with it step by step, ChatGPT can do impressive things. For example, we can start with an outline and flesh out each part with the help of ChatGPT.
- **Editing:** ChatGPT is an excellent editing assistant. It can be used for copy editing, proofreading, rephrasing, and style adjustments.
- **Translation:** ChatGPT can translate very well using dozens of languages. It can also improve the translation by providing context, such as an example of a document in the source and destination languages.
- **Summarizing:** ChatGPT can summarize articles, speeches, and papers. It becomes more accurate when guidelines are provided, such as which topics to highlight.
- Brainstorming Ideas: ChatGPT can provide all kinds of assistance, from suggesting discussion points for a presentation to planning a trip.
- **Writing Code:** ChatGPT is a very good coding assistant and turns functionality descriptions into working code in programming and scripting languages.

9.6 Impact of ChatGPT on Jobs

Job losses anticipated due to ChatGPT include:

- **Customer Support Representatives**
  Basic customer support queries can be managed by ChatGPT, potentially reducing the need for human representatives.
- **Data Entry Clerks**
  ChatGPT can automate data entry tasks, leading to a decrease in demand for human clerks.

- **Content Moderators**
  ChatGPT can assist in moderating content and identifying potential issues, reducing the number of human moderators required.
- **Information Retrieval Roles**
  With ChatGPT's ability to retrieve information quickly, certain roles focused on information retrieval may become less necessary.
- **Transcriptionists**
  ChatGPT can transcribe audio or convert speech to text, which may impact the demand for human transcription services.
- **Basic Research and Analysis**
  ChatGPT's ability to generate and analyze data may reduce the need for some entry-level research and analysis positions.

Figure 9.1 summarizes the impact of ChatGPT on jobs.

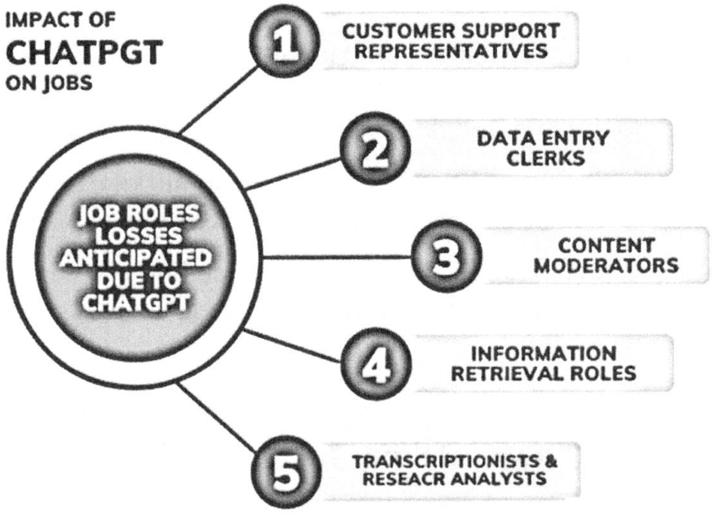

Figure 9.1: Impact of ChatGPT on Jobs

9.7 Conclusion

The rise of generative AI has the potential to be a major game-changer for Shared Services organizations. This technology, which allows for the creation of original content by learning from existing data, has the power to revolutionize industries and transform the way companies operate. By enabling the automation of many tasks that humans previously did, generative AI has the potential to increase efficiency and productivity, reduce costs, and open new opportunities for growth. As such, businesses that can effectively leverage the technology are likely to gain a significant competitive advantage. ChatGPT's knowledge is limited to the data it is trained on. Therefore, it cannot answer questions about all events. One workaround is to provide it with context by copying and pasting the text of a recent report before we prompt and instruct ChatGPT to condition its output on the text of the report.

CHAPTER 10: INTELLIGENT AUTOMATION TOOLS

Due to the abundance of choice, it is challenging to distinguish between the different types of Intelligent Automation tools. These tools can be classified as Artificial Intelligence, Process Intelligence, API-based, and UI-based Automation tools. Figure 10.1 summarizes the Intelligent Automation tools (Dilmegani, 2023).

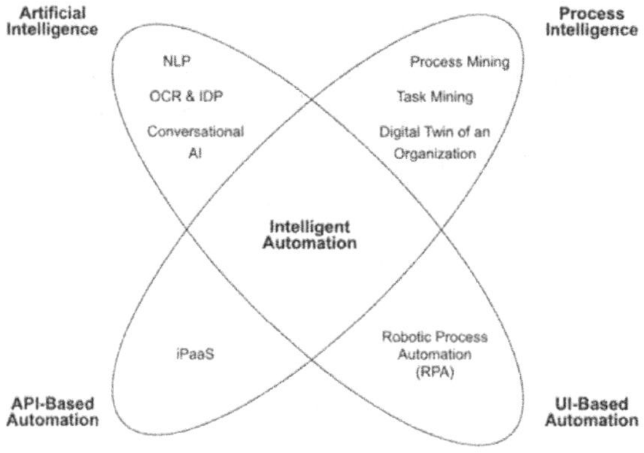

Figure 10.1: Intelligent Automation Tools

10.1 Key Capabilities of Intelligent Automation Tools

The features that intelligent automation tools provide and explore which tools provide for end-to-end process automation are summarized as follows:

1. **(AI) Capabilities**

Natural language processing (NLP)

- NLP is a branch of artificial intelligence that deals with the understanding and manipulation of human language in text or voice form. By leveraging NLP algorithms, intelligent automation tools can:
- Read and interpret unstructured data coming from various sources such as emails, social media, customer service conversations, or invoices,
- Understand the context of the document and extract relevant information from them,
- Classify, summarize, or validate extracted information.

2. **Optical Character Recognition (OCR) and Intelligent Document Processing (IDP)**

Optical Character Recognition (OCR) is a technology that allows computers to read text from images. There are OCR tools in the market, but traditional OCR is limited in its ability to accurately recognize text, particularly in handwritten or poorly scanned documents.

IDP, or intelligent document processing, is a newer technology that uses OCR and artificial intelligence methods such as NLP and computer vision to process documents. IDP systems are designed to:

- Understand the structure of a non-machine-readable document and the meaning of its content,
- Reduce noise in the documents,
- Accurately recognize handwritten or poorly printed documents.

These features allow intelligent automation tools to convert unstructured documents, such as images or PDFs, to a machine-readable format and extract information from them.

3. **Conversational AI**

Conversational AI and chatbots have become increasingly popular tools for businesses, as nearly 80% of CEOs have changed or want to change how they manage client engagement using conversational AI technologies. In addition to their use cases in customer services, intelligent automation tools with conversational AI capabilities can improve the employee experience by making it easier for employees to access information and complete tasks. For example, an intelligent bot can be used to answer common questions that employees have about the company's policies. The chatbot can then direct the employee to the appropriate resource or even complete the task on the employee's behalf. This can free up employees' time so that they can focus on more important tasks.

4. Process Intelligence Capabilities

Process Mining

- As businesses continue to strive for process optimization, process mining has emerged as a valuable tool for understanding and improving processes. Process mining uses real-time data and incident records to:
- Understand the as-is state of existing business processes rather than the ideal ones,
- Identify process bottlenecks and inefficiencies,
- Present process improvement opportunities.

By combining process mining with RPA, intelligent automation tools allow businesses not only to identify process improvements but also automate them. Using process mining in conjunction with RPA can increase the business value by 40% while reducing RPA implementation time by 50% and RPA project risk by 60%.

5. Task Mining

Task mining is another process intelligence method that allows organizations to understand how they perform their tasks through user interaction data. While process mining focuses on how processes are handled, task mining focuses on specific steps or tasks within a process.

Since they are closely related technologies, most process mining solutions also provide task mining capabilities. For example, if process mining reveals that a particular process is taking longer than necessary, task mining can be used to pinpoint which steps in the process are taking up the most time.

Intelligent automation tools with task mining capabilities can allow businesses to understand how users interact with process applications so that intelligent bots can be configured to automate these tasks. This approach can improve process efficiency and effectiveness by reducing process variation and eliminating the need for manual intervention.

6. Digital Twin of an Organization (DTO)

A digital twin of an organization (DTO) is a comprehensive digital model that captures all the data, processes, and interactions within an organization. A DTO provides real-time visibility over business processes and assets as it:

- Analyze individual processes and interactions between them,
- Identify inefficiencies,
- Run simulations about the outcomes of intended changes.

A DTO can be used with process and task mining to identify opportunities for intelligent automation. For more, check out our article on how DTOs and process mining complement each other.

7. UI-based Automation Capabilities

Robotic process automation (RPA)

Robotic process automation (RPA) is the core technology of intelligent automation tools. RPA tools have been on the market since the early 2000s, and traditional RPA tools are based on screen scraping to create bots that can interact with GUI elements to complete repetitive, well-defined, and rules-based tasks.

Intelligent automation integrates RPA with AI and other capabilities to:

- Automate complex tasks with unstructured information,
- Handle exceptions,
- Make data-driven decisions.

Most leading RPA vendors provide either built-in cognitive capabilities or enable developers to add these capabilities to create intelligent bots. Feel free to check our article on the difference between intelligent automation and RPA for more information.

8. API-based Automation Capabilities

iPaaS

An integration platform as a service (iPaaS) provides tools, such as connectors, adaptors, APIs, etc., to enable different software applications to share data and connect with each other. For example, through APIs provided by iPaaS, an e-commerce website can connect with a CRM system and create new records or retrieve customer information.

Intelligent automation tools with API-based automation capabilities can automate tasks requiring data from multiple software applications, including modern cloud and legacy applications.

10.2 Criteria for Selecting Intelligent Automation Tools

Since intelligent automation is built upon RPA and intelligent automation tools are provided by RPA vendors, the process of selecting an intelligent automation tool is like selecting an RPA tool. Some criteria to consider when selecting an intelligent automation tool include:

- **Programming Options**: There are four main types of programming intelligent automation bots: code-based, low-code/no-code, recording-based, and self-learning bots.
- **Attended vs. Unattended Bots**: Attended Bots, also called robotic desktop automation (RDA) or human-in-the-loop automation, reside on the user's machine and are triggered by them to complete tasks. Unattended Bots complete tasks autonomously in the background. Most intelligent automation vendors enable users to create both types of bots.
- **Pricing**: Pricing models of intelligent automation vendors can be complex and lack transparency.

10.3 Conclusion

Intelligent automation can revolutionize business operations by combining automation technologies and Artificial Intelligence to improve efficiency, save costs, and enhance accuracy. Data shows almost half of businesses use automation in some way to reduce errors and speed up manual work. Intelligent automation takes this trend to the next level. It is essential for businesses to understand its definition and various applications as it becomes table stakes for companies worldwide.

Intelligent automation tools and software use technologies such as Artificial Intelligence and Robotic Process Automation to automate mundane, repetitive tasks. Intelligent automation tools aim to streamline processes, reduce costs, and improve the efficiency of business operations.

CHAPTER 11: USING INTELLIGENT AUTOMATION TO TRANSFORM SHARED SERVICES

Shared Services organizations and their internal customers face significant challenges as operations continue to grow in complexity and cost. This is often due to organizations managing an ever-increasing portfolio of applications and volume of data, all while maintaining compliance and enabling missions. To manage these challenges effectively, Shared Services organizations need solutions that operate on the principles with which service delivery models were created. Digital automation tools such as Artificial Intelligence and Robotic Process Automation meet the service delivery principles. Digital technologies have matured to a point where these can transform service delivery models and operations. Shared Services organizations can now use digital automation tools to integrate disparate and disconnected systems and automate processes. These tools can act as a catalyst to help streamline processes, improve data management, reduce costs, improve customer experience, strengthen controls, and reduce backlogs.

Intelligent automation combines methods involving people, organizations, and technologies involving machine learning. Intelligent automation is aimed at automating end-to-end business processes on computers. It delivers business outcomes on behalf of the employees, collaborating with them to deliver faster, better, and cheaper services. This improves not only the employee experience but also the customer experience. Intelligent Automation helps employees work faster and better and gives them more time to focus on what really matters. For example, insights, comments, relationships, and managing their teams because intelligent automation is working for them on the most transactional and repetitive activities.

We always say, 'I don't have the time to do that because I need to do all this'. Now, we have more time for qualitative work. The employee experience is critical, but so is the customer experience. Waiting in a queue to get a train ticket or seeing a bank agent- all of this can be solved by intelligent automation. More than 90% of customers do not even bother to complain, and 90% of them will never return. So, it is really a black box here to understand what makes a client satisfied and how to get their loyalty - and intelligent automation is in the middle of this (Bornet, 2021).

11.1 Digitally Enabled Operating Model

Digital automation technologies deployed simultaneously across multiple groups or processes without coordination can result in varying development standards, approval and management protocols, and review criteria, making it difficult to manage effectively. To help prevent this scenario, Shared Services organizations should incorporate automation tools into planning, strategy, and operating model development. A strategic operating model can enable Shared Services organizations to understand their own systems and capabilities, enabling a targeted deployment. By developing an understanding of current systems—including system providers, technical environment, and organizational alignment—and their respective capabilities, SSOs can select groups and processes where these tools can make the greatest impact. Organizations that strategically plan for the integration of automation tools into their operating models will be able to maximize the demonstrated benefits and be rewarded with smoother implementations.

As Shared Services organizations reach digital maturity, these organizations are increasingly going beyond transactional activities to provide things like higher service quality and advanced data analytics while appointing global process owners to standardize processes like procure-to-pay (P2P) and order-to-cash (O2C). All companies around the world are performing an end-to-end P2P process, which is about procuring goods and services from vendors. Intelligent automation will help, first, to select the vendors using machine learning. Then, it is about sending orders to those vendors and receiving and processing invoices after the services or goods have been received - and this can be done using natural language processing. And finally, the payment to those vendors. It is quite systematic and can be done using robotic process automation.

The push for quality has also placed a premium on higher-skilled resources, marking a shift away from the offshore Shared Services model. As companies attempt to reposition their Shared Services organizations as centers of excellence, the automation of processes such as Order-to-Cash and Procure-to-Pay is emerging as the perfect complementary solution. Not only does it facilitate traditional objectives of Shared Services (e.g., time savings, cost reduction, etc.), but it is also proven to be instrumental in helping companies shape and achieve their larger digital transformation goals. By giving hours back to the business, leaders can then decide where to invest that time to deliver the most value (Esker, 2018).

11.2 Innovative Service Delivery Models

In the past, Shared Services organizations focused on Enterprise Resource Planning (ERP) system implementations. These typically require significant planning, training, and the reassignment of resources and processes within the future-state environment. The advent of robotics, cognitive, and AI, which are more agile technologies, can enable mission support and Shared Services organizations looking to modernize their IT infrastructure to improve existing ERP systems and augment legacy operations. These new tools can help bridge system gaps or shortcomings without the need for an extensive, complex, time and resource-intensive system overhaul. Companies often implement a variety of applications to supplement primary ERP systems that do not directly communicate or interact with each other. This can be a significant pain point for companies, as disjointed applications and systems often result in process workarounds and added manual efforts. Digital automation, however, can be the connective tissue to bring applications together into a single, streamlined process. Implementing these enhancements in a targeted manner can help organizations maximize the efficiency and capability of existing systems and applications cost-effectively.

Automation can also be deployed to enable system modernization efforts. Organizations preparing for a large system implementation can use automation to ease and minimize the time to go live by performing time-intensive preparatory activities, such as data cleansing and extraction. These tools are nimble and can be utilized across a variety of systems so that existing tools can be re-configured for use even within a newly implemented system.

Digital automation technologies deployed simultaneously across multiple groups or processes without coordination can result in varying development standards, approval and management protocols, and review criteria, making it challenging to manage effectively. To help prevent this scenario, Shared Services organizations should incorporate automation tools into planning, strategy, and operating model development (Clarke, Shared Services Digital Automation Tools, 2021).

A strategic operating model can enable SSOs to understand their own systems and capabilities, enabling targeted deployment. By developing an understanding of current systems—including system providers, technical environment, and organizational alignment—and their respective capabilities, SSOs can select groups and processes where these tools can make the greatest impact. Organizations that strategically plan for the integration of automation tools into their operating models will be able to maximize the demonstrated benefits and be rewarded with smoother implementations.

11.3 Technologies Powering Transformation Solutions

It is really about a constructive collaboration between what the people do and will do and what the machines do and will do. Machines are supported by machine learning and robotics technologies and can perform any transactional, repetitive tasks very well. But they need to be trained on vast amounts of data, so it means those tasks need to have been repeated in the past many, many, many times. But machines are not good and will not be good at all for at least the next few decades in tasks such as creativity, relationship building, critical thinking, or anything involving social skills. So, it is really about humans focusing on those capabilities that make us different from technology. Some of the notable technologies powering transformation solutions include:

- **Robotic Process Automation (RPA)**

  RPA is software that can be programmed across applications to do basic human tasks (e.g., file handling, data entry, retrieving orders or submitting AP invoices in a portal). Unlike AI, RPA is limited in that it cannot train itself and make "informed" decisions, which is why a combination of the two is ideal.

- **Process Automation**

  Process automation is a term that describes a set of intelligent technologies built on APIs that allow holistic system integration across the enterprise.

- **Machine Learning**

  Machine learning is a type of Artificial Intelligence (AI) that uses algorithms to automatically "learn" how you manage orders, process invoices, and handle exceptions of all kinds. Notably, machine learning can continue to improve its understanding of data over time, optimizing accuracy and efficiency. About 1 in 3 Shared Services Centers are using AI and machine learning.

- **Deep Learning**

  Deep learning is a subset of machine learning that can automatically perform complex tasks. For example, a solution can automatically and accurately extract data from orders and invoices, even if it's the first time the solution sees it.

11.4 Transformational Leadership and Transactional Leadership

The proliferation of leadership models can leave practitioners confused about which model to use in guiding their leadership development initiatives. Although 'new' leadership models (e.g., authentic, ethical, and servant leadership) suggest theoretical differences, considerable overlap has been shown between these models and longer-standing ones such as transformational leadership. Major leadership models highlight evidence for empirical redundancy between new leadership models and transformational leadership and discuss meta-analytic findings between transformational leadership and outcomes of leadership. These leadership models add little incremental validity beyond transformational leadership in predicting various leadership outcomes. Moreover, transformational leadership demonstrates medium to large effect sizes on a range of individual, team, and organizational outcomes. Taken together, it is suggested that organizations can benefit by focusing their resources on transformational leadership development.

Transformational leadership is a leadership style that can inspire positive changes in those who follow. Transformational leaders are generally energetic, enthusiastic, and passionate. Not only are these leaders concerned and involved in the process, but they are also focused on helping every member of the group succeed. Transformational leaders take control of situations by conveying a clear vision of the group's goals. These leaders have a marked passion for the work and an ability to make the rest of the group feel recharged and energized.

Transformational leadership empowers people to accomplish positive change through big vision, inspiration, and a call to action. Leaders enable change by emotionally connecting with their audience and inspiring them to achieve something greater than themselves. By communicating a positive vision of the future, they motivate and encourage their followers to keep pushing forward. The result is collective action that leads to innovation, massive transformation, and the betterment of society.

Transformational leaders focus on helping members of the group support one another and provide them with the support, guidance, and inspiration they need to work hard, perform well, and stay loyal to the group. The primary goals of transformational leadership are to inspire growth, promote loyalty, and instill confidence in group members. The four main elements of transformational leadership are (Cherry, 2023):

1. **Intellectual Stimulation:** Transformational leaders not only challenge the status quo; they also encourage creativity among followers. The leader encourages followers to explore new ways of doing things and new opportunities to learn.
2. **Individualized Consideration:** Transformational leadership also involves supporting and encouraging individual followers. To foster supportive relationships, transformational leaders keep lines of communication open so that followers feel free to share ideas and that leaders can immediately recognize each follower's unique contributions.
3. **Inspirational Motivation:** Transformational leaders have a clear vision that they can articulate to followers. These leaders can also help followers experience the same passion and motivation to fulfill these goals.
4. **Idealized Influence:** The transformational leader serves as a role model for followers. Because followers trust and respect the leader, they emulate this individual and internalize their ideals.

Transformational leaders inspire people to achieve extraordinary results. Group members are not just encouraged to participate; they are also inspired to become leaders themselves. Transformational leaders can accomplish this by being responsive and empowering. The individual's goals become better aligned with the purposes of the entire group so that each group member's success furthers the organization's goals. This style of leadership can have a positive effect on the group. Some of these effects include:

- **Better performance**: Those led by transformational individuals have better performance and are more satisfied than those in groups with different types of leaders.
- **Better well-being**: Employees who identified a higher level of transformational leadership in their employers also reported higher levels of well-being. The effect stayed significant even after researchers controlled for factors linked to well-being, such as job strain, education, and age.
- **Sense of empowerment**: This is attributed to transformational leaders believing that their followers can do their best, leading group members to feel inspired and empowered.

Becoming a more transformational leader may provide many benefits. Such results can help companies develop leadership training programs that can be used to teach transformational leadership skills. Acquiring communication skills such as resolving workplace conflicts and recognizing employees' needs are important parts of transformational leadership. Such programs can serve as essential parts of health promotion efforts in the workplace to help improve employee well-being. Leadership experts suggest that a robust and positive future vision plays a critical role. Transformational skills can be developed by being:

- Genuine
- Passionate
- Supportive
- Trustworthy

Transformational leadership is often contrasted with transactional leadership, which is a style that is essentially the opposite of transformational. In the transactional approach, leaders motivate their employees using rewards and punishment. While the transformational approach focuses on communication, inspiration, and positive reinforcement, transactional leaders supervise, monitor performance, and create routines to maximize efficiency. While the transactional approach can be effective in some situations, it does not foster loyalty, innovation, or creativity. While transformational leadership is often viewed as one of the best approaches to leadership, that doesn't mean that it is necessarily right for every situation. For example, when group members need more guidance and direction, utilizing a more transactional approach can be more effective.

The transactional style can help improve group cohesion and commitment, but it can also contribute to burnout when group members feel constantly pressured to give up all their time and effort to support the goals of the group. In situations where a lot of creativity and innovation are important to success, a transformational style is often a beneficial approach. However, if the focus is on achieving a prescribed set of short-term goals, taking a more transactional approach might lead to less chaos and better results.

11.5 Benefits of End-to-End Automation

The benefits of End-to-End automation can be summarized as follows:

1. **Universal Simplicity**

   Whether we are grouping AP, AR, or customer service activities, the Shared Services organizations must deal with issues caused by consolidation. Chief among them includes navigating the disparate ERPs implemented across various subsidiaries along with a lack of visibility throughout processes that results in different levels of data and an inability to get accurate forecasting. Best-in-class automation solutions overcome this challenge by integrating with any existing ERP system, often with standard connectors for the most pervasive ERP providers. What's more, automation solutions promote visibility across all formats and reception channels, whether it's orders or invoices — including EDI. It all equates to better business continuity, consistency, and collaboration.

2. **Improved Compliance**

   Rolling services into Shared Services is an effective way to ensure business activities are performed according to internal and external rules. However, unaddressed manual processes can still leave the company vulnerable. Automation is a great option for organizations concerned with compliance, especially those looking to expand their coverage area.

3. **Greater Control & Security**

   Maintaining compliance and keeping the business online and secure is an essential component to ensuring Shared Services' success. When it comes to automation solutions, look for providers that can offer:

- Data centers with physical access to environmental & electrical security.
- Freedom to modify or revoke the rights of existing users or even give them an administrator profile.
- Single sign-on functionality & data transfer by secure protocol.
- 24/7 year-round worldwide technical support service & continuous availability.

4. **Enhanced Visibility**

When it comes to Shared Services, we can't control what we can't measure. Deeper, more agile analytics and reporting are essential for future success. Traditional manual processes can make it difficult to access and/or analyze data. As a result, time, money, and opportunities are wasted trying to obtain an accurate view of enterprise processes. One of the greatest values that AI-driven automation can provide is visibility. The right solution can help a company measure both day-to-day and long-term performance while expanding the scope of its Shared Services organization and achieving added value beyond cost optimization.

5. **Time and Cost Savings**

The competitive landscape leaves little margin left to compete on price alone. However, that doesn't mean that the concept of cost reduction is any less important. Time and cost savings are a primary goal of Shared Services organizations. Tools such as AI and process automation can help achieve these goals by acting as a virtual assistant for each of the employees, eliminating repetitive, transactional tasks and allowing them to focus on activities that drive the greatest business value. Below is a specific scenario for Accounts Payable detailing how automation fosters a more productive environment:

- **Intelligent Data Capture**

  Invoices are captured automatically regardless of the channel on which they arrive. Machine learning then extracts invoice data to be populated into a validation form or auto-approved when no exception is detected.

- **3-Way Match**

  Invoice data is auto-matched with the corresponding orders and goods receipts, enabling staff to quickly review and resolve exceptions from a single, consolidated view.

- **Tracking & Archiving**

  Tracking invoice submission and approval status provides visibility on accrual amounts if invoices haven't been approved. Validated invoices are entered in the accounting book, archived, and made accessible as long as necessary.

- **Rule-Based Workflow**

  Invoices that require payment approvals are automatically routed to the right individual or group in charge.

6. **Agility & Scalability**

   For Shared Services organizations, the ability to ramp up or down quickly based on changes in requirements is imperative. Seek out automation solution providers that can provide robust capabilities designed for ultimate agility and scalability.

7. **Support for Shared Services Expansion**

   The desire to leverage automation across an increasingly broader scale has led many Shared Services organizations to pursue Global Business Services (GBS) and even accelerate its implementation. Automation seamlessly supports globalization for companies ready to make the switch to GBS.

8. **Better Communication**

   In a manual environment, managing customer and supplier relationships can be a challenging and time-consuming endeavor, whether it is answering the phone for a status update on orders or invoices or fielding multiple phone calls to resolve disputes. With the right automation solution in place, however, businesses can foster an environment that is more transparent and mutually beneficial to the business and its customers and suppliers. For some companies, online portals are great options for giving customers and suppliers convenient self-service access to:

   - Manage payment and/or order information.
   - Instantly track and/or submit documents.
   - Perform other process-specific actions.
   - Resolve potential issues via a chat tool and even keep track of what is said.

9. **Employee Satisfaction**

   With Millennials already the largest population of workers, a large part of the increased focus on quality in Shared Services includes attracting and retaining young talent. Automated solutions give

Shared Services organizations a proven digital transformation tool that appeals to Millennials. Most Millennials embrace the value of technology and don't share the same fears of being replaced by bots. They believe that automation will give them more time to spend on creative and value-added activities.

11.6 Conclusion

The value of Intelligent Automation (IA) across industries is unmistakable. With the automation of repetitive tasks through IA, businesses can reduce their costs as well as establish more consistency within their workflows. Digital transformation efforts have been expedited, fueling more investment within the infrastructure to support automation. As remote work also surges, roles will continue to evolve. Individuals focused on low-level work will be reallocated to implement and scale these solutions as well as other higher-level tasks. Middle managers will need to shift their focus on the more human elements of their job to sustain motivation within the workforce. Automation will expose skills gaps within the workforce, and employees will need to adapt to their continuously changing work environments. Middle management can also support these transitions in a way that mitigates anxiety to ensure that employees remain resilient through these periods of change. Intelligent automation is undoubtedly the future of work, and companies that forgo adoption will find it difficult to remain competitive in their respective markets.

The trend is clear that Shared Services organizations are moving up the value chain; innovation, standardization, analysis, and digitization are the top priorities for Shared Services leaders looking to unlock new efficiency improvements and make their organization a strategic center of excellence. Automation solutions powered by AI and RPA technologies have the ability to complement any Shared Services environment, address broad, end-to-end objectives, and allow Shared Services leaders to align their ideas and ambitions with reality. They should be an essential component of any business evolving toward digital transformation.

In the past, Shared Services organizations focused on Enterprise Resource Planning (ERP) system implementations. These implementations required significant planning, training, and the reassignment of resources and processes within the future-state environment. Shared Services organizations can modernize their IT infrastructure to improve existing ERP systems and augment legacy operations using new, more agile technologies like Artificial Intelligence and Robotic Process Automation. These transformation tools can help bridge system gaps without the need for an extensive, complex, time-consuming, and resource-intensive system overhaul.

Companies often implement a variety of applications to supplement primary ERP systems that do not directly communicate or interact with each other. This can be a significant pain point for Shared Services organizations, as disjointed applications and systems often result in process workarounds and added manual efforts. Digital automation tools can bring applications together into a single, streamlined process. Implementing these enhancements in a targeted manner can help Shared Services organizations maximize the efficiency and capability of existing systems and applications in a cost-effective manner. Automation tools can also be deployed to enable system modernization efforts. Shared Services organizations preparing for a large system implementation can use digital automation tools to ease and minimize the time to go live by performing activities such as data cleansing and extraction. These tools are nimble and can be utilized across a variety of systems.

CHAPTER 12: INTELLIGENT AUTOMATION TOOLS NEEDED FOR SHARED SERVICES

Business Support functions and Shared Services organizations are under increasing pressure. Not only do they have to hit their service targets while finding new cost-out opportunities, but they also must do all this under an unprecedented and ever-growing volume of digital communications - emails, service requests, and support tickets. To date, many intelligent automation initiatives have failed to make a real impact, with agents trapped in a cycle of falling response times, client frustration, and operational slowdowns (Challis, 2021).

When Shared Services organizations aren't running efficiently, the whole business suffers. We need to be aware of this as we seek to enhance the capability of Shared Services organizations through intelligent automation and hyper-automation. While talk of self-healing bots, autonomous everything, and automated enterprise is everywhere, there are technologies that are already technically feasible and deployed.

12.1 Key Technologies Required in the Future

Intelligent automation (IA) combines Robotic Process Automation (RPA) with Artificial Intelligence (AI) and other technologies to create workflows that don't just function automatically but also think, learn, and improve without human intervention. Essentially, IA takes automation and gives it superpowers with the help of AI and other advanced cognitive technologies. It's all about making the robots smarter so they can do the thinking for you and get things done quickly. Simply put, RPA is a less intelligent IA. RPA replaces manual and repetitive work using automation tools like bots. IA introduces cognitive technologies like AI and computer vision into the mix to automate processes that formerly required human thought.

For example, a bot that automatically categorizes users in a CRM based on how they subscribed to a newsletter is a form of RPA. IA might involve using subscriber interaction data (clicks, bounce rate, etc.) to add suggestions to a company's CRM, informing future newsletter content. The following key technologies will supercharge automation in the future:

1. **Conversational Data Intelligence**

   One of the largest challenges facing Shared Services - on top of ever-growing request volumes and the shift to hybrid working - is the lack of insight into the forces driving these shifts. To plan and resource effectively, it's vital that shared services and transformation leaders understand which requests are creating the most work and what the problems are that slow business-critical processes. It's critical they understand not only *how* work is getting done but *why* workflows are being triggered in the first place. This helps them resolve issues at the source rather than just treating the symptoms periodically.

   However, much of this insight is hidden in unstructured communications data - primarily within emails - that can't be analyzed or automated by most traditional intelligent automation tools. Being able to process and understand all this unstructured data at the required scale and speed demands strong natural language processing (NLP) and analytical capabilities, which thankfully can be found in conversational data intelligence tools.

   Conversational Data Intelligence platforms combine NLP and machine learning to rapidly train models that can accurately and reliably convert freeform natural language into structured data that are ready for analysis and automation. Shared services and transformation leaders then have the insight and data they need to scale automation effectively and drive improvements that deliver real ROI. Data extraction from email using NLP is a key investment priority.

2. **Process Mining & Task Mining**

   A suite of automated process discovery tools is making waves by helping organizations understand processes in unprecedented detail. Though not purely focused on automation opportunities alone, these technologies will provide insight into process efficiency that's hard to match.

   Process Mining and Task Mining are technology platforms that can enhance process discovery and improvement initiatives. Automated process discovery is like leaning on steroids. Using system, click, and user behavior data, these tools show the reality of processes rather than a picture distorted by the opinions of subject matter experts. This gives you a baseline to understand the work your service agents are completing while also revealing a pathway for future process improvements and automation.

3. **Optical Character Recognition & Intelligent Document Process**

Optical character recognition (OCR) and intelligent document process (IDP) have seen a huge usage boost thanks to intelligent automation's growing focus on RPA. Adding document understanding to your shared service capability provides a whole new area of work to be structured, digitized, and, therefore, automated by RPA. If semi-structured documents, such as invoices and application forms, are now fair game thanks to OCR and IDP technologies, organizations gain easy access to masses of new structured data sources.

4. **Robotic Process Automation**

Intelligent automation makes use of many complementary technologies, one of the most popular and widespread being Robotic Process Automation (RPA). Although the core capability of RPA is to mimic human workers and replicate their keystroke activity, its far-reaching application - from delivering a sales order into the Customer Relationship Management (CRM) system to processing a customer account amendment or onboarding a new employee - makes RPA one of the most useful software technologies available. Pairing RPA with the prior three technologies can supercharge your automation value-creation initiatives.

Companies should not focus too heavily on the trends that are garnering the most attention. By focusing on only the most hyped trends, they may miss the significant value potential of other technologies and hinder the chance for purposeful capability building. Instead, companies seeking longer-term growth should focus on a portfolio-oriented investment across the technology trends most important to their business. By carefully assessing the evolving landscape and considering a balanced approach, companies can capitalize on both established and emerging technologies to propel innovation and achieve sustainable growth.

12.2 Key Features of Automation Technologies

Here is a checklist of key features of automation technologies. There is a need to determine which of these features are required in the automation software and which ones can continue to be used through the integrated CRM and ERP systems (PlanView, 2021).

1. **Lifecycle Features**

- Visibility into all key customer lifecycle information in a way where everything makes sense.
- Contact documentation tools so that you never lose sight of who your key customer contacts are, so you can reach them when needed.
- Quote-to-cash project lifecycle management tools to discover every aspect of the customer lifecycle and capitalize on every financial opportunity.
- Project summaries to immediately know your customers' lifecycle health based on the most important KPIs.

2. **Contract Management Features**

- Customizable contract definition tools to establish financial and billing rules and rates that make up the contractual agreement to drive delivery success.
- Document sharing to attach proposals, contracts, NDAs, and other documents for greater knowledge sharing.
- Activity trackers to monitor actual performance against prospects, leads, and pipeline opportunities and stay aligned with customer expectation

3. **Opportunity and Delivery Features**

Forecasting tools to gain detailed financial visibility into specific contractual expectations for visibility into future resource needs.

- Budgeting tools to build budgets and specify the services, products, and expenses anticipated on a project, as well as the ability to support multiple rates on the same resource on the same project.
- Configurable workflows to establish the structure and cadence required for consistent, profitable delivery.

4. **Project Management Features**

- Integrated task management tools to create, edit, and manage tasks and assignments with financial flexibility and greater control.
- Process workflows that leverage the power of intelligent automation to drive project consistency and governance.
- Project templates that allow you to hit the ground running – in little to no time.
- Collaboration tools that allow your teams to collaborate with each other and customers on their work, whether remote or on-site.

5. **Resource Management Features**

- Skills and expertise-based views to align resources to the right work.
- Dynamic Gantt charts that provide a timeline distribution of all availability impacting your dedicated resources.
- Access control to restrict access to a subset of resources within a particular workgroup as needed.
- Quick search capacity to rapidly locate resources based on various parameters like assigned work, qualifications, and more

6. **Financial Management Features**

- Financial visibility tools to gain real-time insights into forecasted and actual revenue and variances to ensure overall financial health.
- Revenue recognition tools to accurately track recognition revenue based on project or milestone completion.
- Time and expense tracking tools that provide anytime, anywhere access to time and expense tracking and approvals for timely invoicing.
- Budgeting tools that ensure your projects are tasked and staged appropriately to meet your financial objectives.
  - Multi-currency and taxation options to bill and recognize revenue in the appropriate currency for a project while complying with tax standards.

7. **Analytics Features**

- Interactive dashboards that use out-of-the-box, preconfigured visualizations with actionable insights.
- Explorations that allow you to dive deep into relevant trends that the system automatically provides for you – all on its own.
- AI and natural language details that enable you to input questions about data and get contextual responses

12.3 Impact of Intelligent Automation Technologies on Shared Services and Global Business Services (GBS)

The increasingly digitized and globalized world economy driven by innovation, disruptive technologies, and interconnectivity has transformed the traditional business models of Shared Services and Global Business Services organizations. These organizations are continuously evolving to support more complex core business processes and enable more agile, flexible, sustainable, and cost-effective operations. The digital ecosystem, driven by economic viability and technology maturity, has emerged as a key prerequisite and core competency to drive enterprise growth and achieve successful business outcomes for service organizations. Intelligent Automation, one of the key components of the digital ecosystem, has enabled the transformation of Shared Services organizations by providing a rapid return on investments, non-invasive modernization, and key business and operational process improvements. Automation and machines have become fundamental for performing tedious, mundane, and repetitive tasks to deliver data-driven and cost-effective insights.

To understand the potential impact of intelligent automation technologies on Shared Services, let's look at some of the decisions that were made during the implementation of Shared Services (Dement, Manning, & Robinson, 2023).

- **Business Process Outsourcing (BPO) Decisions**

  For years, companies have turned to BPO solution providers to lower the costs associated with standard, rule-based, repetitive processes. BPO solution providers, in turn, moved their service centers to developing countries where labor costs were lower. But what if the processes involved could be automated via RPA bots? This would either reduce the need to outsource selected processes or would impact the decision as to which BPO service provider to use, insofar as many providers are working to adapt the new intelligent automation technologies to their solution offerings.

- **ERP Decisions**

  RPA's ability to automate repetitive process transactions has multiple ERP implications. For example, one of the daunting tasks facing Shared Services and GBS executives is whether to move to a single-instance ERP solution or deal with the headaches of multiple ERP solutions across their global footprint. RPA solutions can sidestep this decision and serve as an unsophisticated "integration" tool. If your customer service representatives are currently logging onto one system and then using that data to input into another system, it may be cheaper to automate this activity via RPA rather than move to a single ERP solution.

- **Support Service Location Decisions**

  Typically, Shared Services and GBS support hubs have been in specific countries or key regions to utilize local language employees and provide service availability in local time zones. But what if Virtual Assistants could handle a significant portion of the call volume, immediately translate any language, and do this on a 24/7 basis? Country and regional service hubs could be located closer to the company's headquarters where technology solutions may be easier to manage, quality control and intellectual property and privacy issues might be better managed, and expertise services would be more readily available.

- **Location Decisions for Expertise Services**

  Expertise services have typically been located close to the company's headquarters, where service specialists are available. In those instances where AI solutions can perform complex analytical tasks, some or many of these tasks can be performed anywhere in the world.

  Services currently located in foreign transaction centers and regional service hubs might be moved back to the company's headquarters or located wherever Robotic Process Automation (RPA), Virtual Agents (VA), and Artificial Intelligence (AI) talent is available.

  Preparing to harness the impact of any disruptive technology is always problematic. Move too soon, and we will be on the "bleeding edge" of technology. If we move too late, we may lose the competitive advantage. The one thing we cannot do is to ignore it. Figure 12.1 summarizes the impact of Intelligent Automation technologies on Shared Services and GBS operations.

Potential Impact of Disruptive Intelligent Automation Technologies (IATs) on GBS

Type of Work	Description	Shared Services Delivery Tier	Applicable IATs	Typical Work Distribution Pre-IATs	Percent Labor Reduction
Strategic	Strategic direction and policy making	N/A	N/A	7%	0%
Specialist/Consultative	Deep specialist focus, complex analysis, decades of experience	Tier 3	VA, AI	13%	5%–30%
Functional/Programmatic	In-depth subject matter expertise; close proximity to customer	Tier 3 Tier 2	VA, AI	27%	15%–40%
Transactional/Administrative	Repetitive tasks, routine data entry, issue resolution	Tier 1 Tier 0	RPA, VA	53%	20%–50%

Figure 12.1: Impact of Intelligent Automation Technologies on Shared Services and GBS

We cannot ignore that these technologies are going to fundamentally affect how Shared Services and GBS organizations deliver their services. Whether they represent a problem or an opportunity, it is better to face them rather than postpone the inevitable. By effectively assessing the opportunities and piloting the technologies, we can increase the likelihood of implementing an effective solution to improve Shared Services and GBS operations.

12.4 Benefits of Intelligent Automation in Shared Services

Most Shared Services organizations have either implemented or are in the process of implementing intelligent automation. By adopting intelligent automation, these organizations aim to deliver greater efficiency and quality to their operations (Suresh, 2022). Benefits of intelligent automation include:

- **Increased Productivity:** Once set up, RPA delivers results immediately as mundane manual processes are replaced by high-performing bots. Tasks like emailing updates, filling out forms, and sourcing information that may take up the entire working hours of employees can be done within a matter of minutes. Research suggests that RPA eliminates nearly 20% of non-value-added tasks. RPA bots operate at high accuracy levels, making the entire process error-free, efficient, and reliable.

- **Enabling Talent:** As labor-intensive tasks are automated, employees are free to manage more complex exceptions. They can be directed to focus on value-added activities that bring greater returns to the organization. They can take up more responsibilities like vendor management, data analysis, or interacting more with customers to enhance relationships and loyalty. With intelligent automation, customers can also make use of self-service portals to get the information they need themselves.

- **Greater Compliance & Cost Savings:** With the right strategies, processes are mapped out and approved for successful automation. So, when that is done, enterprises need not worry about compliance as

RPA does automation to the T. By ensuring compliance, businesses consistently deliver quality as all industry standards and procedures are followed. RPA enterprises see visible gains as operational costs are cut by nearly 30%. Bots can operate 24/7, never getting tired, resulting in greater ROI.

- **Improved Visibility:** By automating a centralized function like a shared services center, enterprises can see unprecedented levels of visibility. With end-to-end automation, a process's entire lifecycle is handled from a central location. With dashboards and real-time analytics, the management has a clear view of all operations and processes to spot and rectify any bottlenecks.
- **Scalable Solution:** The biggest benefit of RPA is that the technology can be overlaid on existing systems. This enables the platform to be compatible with ongoing developments in sophisticated algorithms and machine learning tools. Intelligent automation also allows for instant ramp-ups and downs to match the demand, thus increasing the scalability capabilities of Shared Services organizations. It also significantly reduces the hassle of training and managing new staff. With improved operational agility, enterprises can propel their speed to market and respond faster to market dynamics.

12.5 Success of Intelligent Automation in Shared Services

While many factors outline the success of Intelligent Automation in Shared Services Organizations, it has become increasingly imperative to focus on three main areas - Governance, Process Discovery, and Scalability (Enate, 2021). These are explained in detail below:

- **Governance**

  While many intelligent automation projects in Shared Services start with the automation of repetitive and rule-based manual processes like data entry/matching, they gradually move on to perform more operationally intelligent tasks. Here is when governance plays a key role in ensuring there is trustworthiness in the ecosystem by monitoring SLAs and facilitating compliance management.

  Also, improper governance can expose Shared Services processes to major risks. For example, exposing sensitive employee/vendor data due to automation malfunction or misconfiguration, data entry mistakes, incorrect decision-making on invoice approvals/calculations, etc. Such instances could also hinder regulatory compliance.

- **Process Discovery**

  Discovering ideal processes for automation can be a challenging task, especially in complex Shared Services organizations. Process analysis and documentation can take up to 60-80 % of automation efforts, especially if it is done manually. Also, apart from just discovering mundane and repetitive processes for automation, the industry must focus on how to automate more complex processes as part of the Intelligent Automation roadmap.

  Processes for functions like Finance & Accounting, Human Resources, Customer Service, and Procurement involve coordination and information flow between multiple stakeholders, thus contributing to longer life cycles. Delays and exceptions can happen due to multiple reasons in such processes, even after the implementation of automation technologies like RPA. Here's where data-driven process discovery tools come in handy, which keep monitoring such processes to identify bottlenecks that can be reduced with automation, thus increasing operational efficiencies.

- **Scalability**

  Automating processes in Shared Services through technologies like RPA is an investment-heavy exercise both in terms of human resources and capital investment. And if such an exercise fails to scale, it can lead to significant cost write-offs and delay digital transformation for the organization. Most automation implementations in Shared Services organizations have been struggling to scale up beyond a few bots due to multiple reasons, which are highlighted below:

- Automation Implementations are eventually plagued with ineffective governance, sub-optimal work handoffs between humans and bots, poor exception handling, and incompatibility with multiple bot vendors, all of which lead to delays and reduce the benefits of RPA
- Struggle to integrate the next layer of future-of-work technologies like AI, ML, OCR, and Cognitive into existing workflows
- Trying to improve service efficiency and effectiveness while managing unexpected increases in service volumes

12.6 Conclusion

Managing Shared Services organizations and making them cost-efficient is a challenging task for any medium to large company. Shared Services and Global Business Services are important internal functions that take on high-volume business processes and administrative and back-end operations typically done by multiple departments and executed under a centralized roof to deliver company-wide benefits. Shared Services may include functions from human resources, operations, finance, IT services, and legal and document management of a company. Traditional Shared Services models are hyper-focused on cost optimization. However, many companies are changing their priorities and looking at Shared Services organizations through the lens of growth opportunities.

Shared Services organizations are looking to break away from the perception of being a pure cost-cutting, transaction-oriented system to truly make a strategic contribution in terms of company performance by improving process control, standardization, and optimization. To move up the ladder in the value chain and enable businesses with agile services, Shared Services organizations are looking to adopt automation. Their rule-based and logical operations make them ideal candidates for Intelligent Automation.

Intelligent Automation can create significant value within a Shared Services organization. It offers attractive strategic and tactical solutions to many business problems that arise from legacy architectures and manual business processes. However, automation technologies must be properly implemented, and this is best achieved by putting in place the right framework, such as a Center of Expertise equipped with the correct set of tools and best practices. Shared Services organizations can re-structure their engagements with the help of Intelligent Automation and achieve great returns considering the scalability it brings in the operating model.

CHAPTER 13: TRENDS SHAPING THE FUTURE OF SHARED SERVICES

Trends such as automation, data governance, migration of judgment-based work, and the hybrid workplace have implications for Shared Services organizations. Shared services leaders need to analyze which trends should underpin their strategies. Automation is being adopted successfully for both customer interfaces and functional processes. Chatbots are being used to supplement live agents since the bots can provide quick answers to basic inquiries or help route inquiries to the right place. Many employees often prefer this real-time service option. Shared Services organizations need to select and implement virtual assistant technology with artificial intelligence to automate many aspects of interactions with internal customers and manage their requirements with remarkable success. Shared Services organizations are also using Robotic Process Automation (RPA) to automate routine processes such as employee data management, payroll management, and employee exits.

As Shared Services organizations have evolved, there has been expanded scope, adjustment of service center roles, revised governance structures, and greater use of technology and automation. Shared Services organizations have moved up the value chain to bring on more complex services beyond the original targets for administrative and transactional work. As the scope of work has expanded, skills and staffing levels in Shared Services have had to adapt as well. There is a broader use of specialists to deliver these advanced services, and the required skill sets have expanded to include consultation, critical thinking, and data analytics. From a governance standpoint, more Shared Services organizations report to multifunction leaders or other executives than in the past due to the expansion of Shared Services models.

Shared Services organizations are embracing advanced technology and automation to gain further efficiencies and improve the internal customers' experience with Shared Services. There are several services that are good candidates for Shared Services organizations. Some of the most common include employee and labor relations, talent acquisition, organizational effectiveness, business analytics, and project management services. With talent acquisition, Shared Services can manage many aspects, including sourcing and screening candidates, scheduling interviews, preparing offer letters, and coordinating employee onboarding. Shared Services organizations are well-suited to offer business analytics services and can lead transformation efforts for the company.

13.1 Embracing the Shared Services Trends

Over decades, Shared Services organizations have been an evolving service delivery operating model that supports the company's internal customers. When successfully deployed and operated, Shared Services can provide the benefit of lower costs, improved service levels, and a tighter, more efficient control environment delivered by an internal (as opposed to an externally outsourced) delivery team. Leaders of Shared Services organizations have had to embrace the new trends and the new way of working, including considering how best to benefit from virtual platforms, remote working, flexible hours, etc., and take advantage of the significant opportunities from both the technology landscape and developments (Corpshore, 2023).

Shared Services models tend to be more siloed and, therefore, are less able to leverage scale and be more integrated. This means that there is a lower risk of a global, cross-enterprise impact if one siloed center is, for example, suffering from a shutdown or lots of staff being off sick. Shared Services models, and especially those leveraging Global Business Services and all its elements, have been able to serve internal customers. The Shared Services trends include the following (Searle, 2023):

1. **Automation and Digitalization**

   The pandemic has accelerated many of the shared services trends that were already in place. It has been clearly demonstrated that world-class Shared Services/GBS organizations, with strong operating models and greater automation, are better prepared for challenges and opportunities. Companies are building hyper-resilient operating models, working with selective, expert external partners and outsourcing relationships, leveraging the very latest in digital technologies and Intelligent Automation_(Robotic Process Automation, Machine Learning, Artificial Intelligence, Blockchain, etc.), and embracing remote working.

Robotic Process Automation (RPA), for example, has advanced to the point that it is being implemented and scaled across most Shared Services organizations, at least to some degree, today. Many GBS/ Shared Services organizations have successfully implemented RPA at scale by embracing this new technology, altering their operating models, redesigning their processes, and training their people. To eliminate some of the very frustrating, manually intensive activities that either always existed or were never solved by the implementation of Enterprise Resource Planning (ERP) solutions.

Shared Services/GBS organizations need to develop and accelerate their digital capabilities to achieve world-class digitally-enabled performance levels. Organizations will need to become highly digitized and self-service enabled. To evolve from world-class to digital world-class, Shared Services will need to accelerate the development and adoption of new digital business capabilities that leverage all the data that is gathered and available today, drive better analytics, and develop value-adding business partnerships. Skills needed for success in the new world include analytical skills, technical skills, plus business and commercial skills, especially as we move closer to what some call the "One Office," where functions, teams, technologies, data, and processes are more closely aligned, integrated and inter-related, and supportive of one another to drive the fastest and most valuable business outcomes. Long gone are the days of the "out of sight and out of mind" unintegrated shipping off to far-flung locations of "non-core" processes and activities.

2. **Virtual Shared Services**

Virtual Shared Services have been discussed for a very long time. This was because, with the significant advancement in new technologies (including bandwidth) and online collaboration tools, supported by a robust service management framework and training and team support techniques, virtual Shared Services has become a trend. This is a necessity but has also proved that it is possible and can bring added benefits in terms of efficiency, effectiveness, and scalability.

3. **Marching into Next Generation**

   Next-Generation Shared Services are defined as the next stage in the evolution of Shared Services. Next-Generation Shared Services organizations will embrace the latest technologies, such as Digitalization and Intelligent Automation (including Robotic Process Automation, Machine Learning, Natural Language Processing, Artificial Intelligence, and Blockchain). They will implement more resilient operating models with hybrid workforces of human and digital workers. They will deliver more services, higher up the value chain, for a wider group of internal customers whilst still ensuring that they get the basics right. Next-Generation Shared Services is likely to play a big part in the move towards One Office and will assist with the closer integration of functions, technologies, processes, data, and teams across the business, all with the goal of producing better outcomes for the company, its customers and employees, and other key stakeholders.

13.2 Impact of Technologies on Shared Services

The following technologies would significantly impact Shared Services organizations in the future:

- AI/ML is expected to have the most significant impact on Shared Services organizations in the future. At present, with a few exceptions, Shared Services organizations are only exploring this technology in simple tasks, such as chatbots. This technology is expected to revolutionize Shared Services with decision support and forward-looking data analysis shortly.
- Internet/package-based automation will also have a significant impact.
- The evolution of BPMS/RPA technologies enables viable intelligent process automation, combining redesign, automation, end-to-end process management, and continuous improvement using AI/ML and RPA.
- In the future, Shared Services organizations will use algorithms to get models to support predictive analytics, neural networks, and complex

event processing. This will allow them to be more consultative in supporting the business units.

- Blockchain would enable automatic transactions between parties and the safe and secure storage of data. Shared Services organizations could facilitate and maintain the blockchain to enable users to transact.
- Shared Services organizations could help users migrate to the cloud and make agreements. In this way, it ensures that data is only stored within the country's jurisdiction.

13.3 Principles for the Future of Shared Services

The principles that drove advances in technologies can be directly applied to the Shared Services environment. Shared Service organizations' mission should focus on internal customers' needs and expectations. There must be partnerships between the public and private sectors to foster innovation. Transparent rules must apply in the marketplace to ensure the playing field is competitive. An effective governance model holds providers accountable for delivering effective Shared Services. Funding and flexibility allow the model to be sustainable, while cost savings ensure it is viable long term (Samberg, 2023).

Shared Services organizations will offer internal customers the flexibility to obtain services built on an "As a Service" technology model. Based on the following principles, the service providers will focus on enhancing offerings and providing timely and cost-effective services:

1. **Infrastructure**

   Infrastructure will be provided as a service (IaaS) and will be delivered commonly across internal customers by cloud providers. We know already that this move is driving significant savings.

2. **Platform**

   Platform as a service (PaaS) is going to be provided by the private sector. The private sector is going to own, operate, and, in a sense, lease the software to a provider. The platform will be configured as needed for providers and agencies; it will no longer be customized for each agency. Providers will then use that platform for all their functionality and services.

3. **Data**

   Data needs to be standardized. For data to be portable between service providers and platforms (Oracle, SAP, etc.), data standards will be needed. Data standards will drive this transformation for each shared services functional area (HR, FM, Grants, etc.). In some cases, different pieces of government may own different parts of the data – for example, the Social Security Administration would own your social security number – but all the data would reside in a commonly accessible data warehouse by both agencies and providers.

4. **Provider**

   Given the IaaS, PaaS, and standard data warehouse model, providers can focus on service. The shared service provider now looks more like AT&T or Verizon – they are all able to provide services on the same platforms but in ways in which they can differentiate themselves based on better offerings, prices, and service levels. Agencies would then select a provider based on the service provision, not the software or technology.

5. **Software**

   The software model will be "as a service," meaning that the private sector will own and operate the software for HR, Finance, Travel, and other transactional functions. The advantages of this model include tiered pricing, service level agreements, and greater innovation, among others. Under the new model, the internal customers can focus on customer service, offerings, and price differentiation. Providers can focus on innovation and customer service rather than spending time and money on IT management. The three layers below the provider level will be occurring behind the scenes to the agencies, offered seamlessly as services. Agencies do not have to concern themselves with the interactions between infrastructure, data, and platforms. From an agency perspective, the service delivery will be a simple road from point A to B. Thus, agencies can focus all resources on core mission activities.

13.4 Reshaping Shared Services Modernization

As Shared Services organizations mature and modernize into a new generation of integrated business services, evolving business models are creating more value for the companies. With simultaneous growth in revenue, reduced operating expenditure, and streamlined management, the modernization of Shared Services will radically alter the corporate world. Several trends are driving Shared Services' modernization as companies shift their attention to operating models, process workflows, and talent development. These trends are integration, governance, digital, geography, and sourcing. Breaking the functional silos for better integration is key to delivering high-value services on a company-wide basis in a consistent and cost-competitive manner. The global business services model integrates end-to-end processes, including procure-to-pay, order-to-cash, and record-to-report, to achieve operational efficiency savings. Most of these processes are being integrated with governing metrics and data analytics to monitor performance. This optimizes processes and allows the delivery of ongoing improvements and value to Shared Services organizations. Finance is the most mature Shared Services function that is provided through an integrated, centralized global model. Finance as an integrated service is transparent, enabling executives to view working capital throughout the system. While this data enables strategic management decisions, an end-to-end Finance service also eases transactional activities throughout the business (Gopal, 2023).

Governance dictates both the strategic and operational aspects of a business. Strategic governance addresses scope, service, investment, and price, while operational governance covers daily items related to processes, technology, and people. Many Shared Services organizations are set up and managed as autonomous business units. Many use the global process service level agreements and a single enterprise resource planning system throughout. Artificial Intelligence (AI) tools help manage customer experience and control vendor management to create improved and efficient service. Large data lakes are harnessed to drive KPIs and operational reviews. Advanced analytics provide insights into customers and products and aid business planning.

13.5 Skills Required for Shared Services Professionals in the Future

Regarding the primary employee skills demanded by Shared Services organizations in the future, the top two are critical thinking and collaboration. Technical knowledge of expertise area and problem-solving skills are other required skills. Humans will remain relevant because of critical thinking. That is something which we cannot impart on the hyper-automation solutions. To achieve that, the base must be in the domain knowledge and technical expertise as well as a focus on the big picture. Collaboration with different moving parts and adaptation to change will become critical in determining success.

The primary skills of Shared Services leaders include innovation and creativity, new business opportunities, high-performance team management, new models of work organization, and strategic management. The new leaders will need a strong knowledge of innovation, technology, and creativity to find new business opportunities to improve efficiency, efficacy, and user experience in the business processes. The service scope will expand to a more analytical/high-end portfolio of services. The Shared Services business model will change from FTE-centric to value-delivered. People at Shared Services organizations will be engaged in doing high-end work and hence will be billed at a higher rate. The Shared Services organizations must be more strategic in nature, providing timely, accurate, and controlled information and entering a business partnership relationship with internal customers responsible for managing the end-to-end process activities. The additional analytics provision by the Shared Services organizations has the power to provide a competitive advantage to the company in the marketplace.

13.6 Conclusion

As businesses continue to evolve, so do their operations. Shared Services organizations have become increasingly popular as companies look to streamline their processes and reduce costs. Shared Services refers to a model where multiple business units or departments within a company share support functions such as IT, HR, Finance, and Procurement. This model has been around for decades.

The role of technology in Shared Services cannot be overlooked. Emerging technologies such as artificial intelligence, robotic process automation, and blockchain will transform Shared Services organizations. These technologies will allow more efficient and automated processes, reducing manual effort and costs. For example, AI-powered chatbots can be used to answer employee queries, reducing the need for human intervention.

The concept of Shared Services is evolving to include outsourcing to third-party providers. This trend has been driven by the need for more specialized services and the desire to reduce costs further. Outsourcing Shared Services can also provide greater flexibility, as companies can access these services on-demand. The shift towards a more customer-centric approach will shape the future of Shared Services organizations, which will increasingly focus on providing exceptional customer service to internal customers. This will require a shift in mindset and a focus on continuous improvement to ensure that customer needs are met. The future of Shared Services will also see an increased focus on data analytics. Such organizations will leverage data to gain insights into how processes can be improved and costs reduced. This will require a shift in skillsets, with Shared Services professionals needing to be proficient in data analysis. Finally, the future of Shared Services will see a greater emphasis on sustainability. Companies will increasingly focus on reducing their environmental impact, and Shared Services will play a key role in achieving this objective. This will require Shared Services organizations to adopt sustainable practices and leverage technology to reduce energy consumption and waste.

Integration, governance, digital, geography, and sourcing trends are powering the future of Shared Services. Successful navigation of these forces will recalibrate companies to improve the experience of internal stakeholders. New business models will help respond efficiently to Finance,

M&A, and future divestiture needs. Shared Services' strategy must evolve to increase responsiveness to growth and take advantage of future business models. In conclusion, the future of Shared Services is exciting and full of opportunities. Emerging technologies, outsourcing, a customer-centric approach, data analytics, and sustainability will all shape the future of Shared Services organizations. As companies continue to evolve, Shared Services will play an increasingly critical role in driving efficiency, reducing costs, and delivering exceptional customer service. It is an exciting time to be involved in Shared Services, and one cannot wait to see the developments in the future.

Bibliography

ACCA. (2015). The Robots are Coming: The Implications for Finance Shared Services. Shared Services & Outsourcing Advisory Group.

Automation, A. (2020). What is Intelligent Automation? Retrieved from Automation Anywhere.

Beebe, E. (2023). Facing the Future of Finance with ChatGPT. Shared ServicesWeekly, 4.

Bornet, P. (2021). What is Intelligent Automation? World Economic Forum, 4.

Challis, E. (2021). Four Intelligent Automation tools you need for Shared Services. Re: infer, 3.

Chazey, P. (2023). Creating an Appropriate Ecosystem to Advance Your Organization's Intelligent Automation Journey. SSON Intelligent Automation Week (p. 20). Chicago: SSON IA Week.

Chazey, P. (2023). Unleashing the Power of Generative AI in GBS. SSOW (p. 41). Las Vegas: Shared Services & Outsourcing Week.

Cherry, K. (2023). How Transformational Leadership Can Inspire Others. Verywell Mind, 7.

Clark, T. (2023). 5 Artificial Intelligence Use Cases More Promising Than ChatGPT. Forbes, 3.

Clarke, T. (2021). Shared Services Digital Automation Tools. Deloitte Consulting, 5.

Clarke, T. (2022). Digital automation tools: Helping improve capacity and efficiency in shared services. Deloitte Consulting, 5.

Corpshore. (2023). The Future of Shared Services. Corpshore Solutions, 2.

Cromwell, J., Harvey, J.-F., Hasse, J., & Gardner, H. (2023). Discovering Where ChatGPT Can Create Value for Your Company. Harvard Business Review, 6.

Cybersecurity, C. C. (2023). Generative Artificial Intelligence (AI). Government of Canada, 4.

Dement, B., Manning, S., & Robinson, T. (2023). How Intelligent Automation Will Impact Today's Global Business Services Model. ScottMadden, 7.

Dilmegani, C. (2023). Intelligent Automation Tools: Key Features & Top Vendors [2023]. AI Multiple, 6.

Enate. (2021). Intelligent Automation in Shared Services. Enate APAC, 5.

Esker. (2018). 9 Ways AI and RPA Can Transform Your SharedServices Center. Esker, 18.

Fausti, M. (2015). Old Dogs Learn New Tricks. Alsbridge.

Fedak, V. (2018). Top 10 Most Popular AI Models. DZone.

Frevvoblog. (2022). Automation + Digital Transformation: The Key Unlocking Business Growth. Docuphase Company, 7.

Gartner. (2016). Robotic Process Automation (RPA): From Hype to Reality.

Gopal, V. (2023). Four Powerful Trends that are Reshaping Shared Services Modernization. Tata Consultancy Services, 3.

Heatth, N. (2018). ZDNet. Retrieved from Managing AI and ML in the Enterprise.

Hodge, B. (2016). Robotic Process Automation Security. S .S. O. Network.

Horton, R. (2015). The Robots are Coming. Deloitte Insights.

Howard, C. (2022). Head of Shared Services Leadership Vision: 3 Strategic Actions for Success. Gartner, 9.

IBM. (2020). What is intelligent automation? IBM, 4.

IBM. (2023). What is Artificial Intelligence (AI)? IBM, 6.

Insights, S. (2020). Artificial Intelligence. Retrieved from SaS.

Kofax. (2019). Accelerating Business Value with Intelligent Automation. Forbes Insights, 20.

PlanView. (2021). The Ultimate Gude to Professional Services Automation. Planview, 14.

Samberg, J. (2023). Principles for the Future of Shared Services. IBM Center for The Business of Government, 3.

Scottmadden. (2022). What are Shared Services. Scottmadden Management Consultants, 2.

Searle, P. (2023). Introduction – Embracing the New Shared Services Trends. Chazey Partners, 3.

Smith, M. (2015). A Refresher on Robotic Process Automation. Cognizant.

Suresh, S. (2022). How intelligent automation supercharges your shared service operations. Aspire Systems, 4.

Troitsky, M., & Tay, A. (2022). How AI and Machine Learning are Transforming Finance. Deloitte, 2.

Ulrich, D. (1995). Shared Services: From Vogue to Value. HR, Human Resources Planning, 12.

Upadhyay, M. (2022). Artificial Intelligence for Managers. Toronto: 189.

Willcocks, Leslie P.; Lacity, Mary C. (2015). The IT Function and Robotic Process Automation. The London School of Economics, University of Missouri.

Willcocks, Leslie P.; Lacity, Mary C. (2016). What is Robotic Process Automation? Steve Brookes Publishing.

Made in the USA
Monee, IL
04 November 2026

67382569R00089